THEOLOGY AND SCIENCE AT THE FRONTIERS OF KNOWLEDGE

NUMBER ONE

REALITY AND SCIENTIFIC THEOLOGY

THEOLOGY AND SCIENCE AT THE FRONTIERS OF KNOWLEDGE

1: T. F. Torrance, *Reality and Scientific Theology*.
2: H. B. Nebelsick, *Circles of God, Renaissance, Reformation and the Rise of Science*.
3: Iain Paul, *Science and Theology in Einstein's Perspective*.
4: Alexander Thomson, *Tradition and Authority in Science and Theology*.
5: R. G. Mitchell, *Einstein and Christ, A New Approach to the Defence of the Christian Religion*.
6: W. G. Pollard, *Transcendence and Providence, Reflections of a Physicist and Priest*.

THEOLOGY AND SCIENCE AT THE FRONTIERS OF KNOWLEDGE

GENERAL EDITOR – T. F. TORRANCE

REALITY
AND
SCIENTIFIC THEOLOGY

THOMAS F. TORRANCE

SCOTTISH ACADEMIC PRESS
EDINBURGH
1985

Published in association with the
Center of Theological Enquiry, Princeton
and
The Templeton Foundation
by
SCOTTISH ACADEMIC PRESS
33 Montgomery Street, Edinburgh EH7 5JX

First published 1985

ISBN 0 7073 0429 6

© Thomas F. Torrance, 1985

All rights reserved. No part of this
publication may be reproduced, stored in
a retrieval system, or transmitted in any
form or by any means, electronic, mechanical,
photocopying, recording or otherwise, without
the prior permission of Scottish Academic Press Limited.

British Library Cataloguing in Publication Data

Torrance, Thomas F.
Reality and scientific theology.
1. Religion and science—1946–
215   BL240.2

ISBN 0-7073-0429-6

Printed in Great Britain by
Clark Constable, Edinburgh, London, Melbourne

To
Donald M. Mackinnon
Philosopher and Theologian
in deep appreciation and gratitude

CONTENTS

General Foreword — ix

Preface — xi

Chapter
1. Classical and Modern Attitudes of Mind — 1
2. The Status of Natural Theology — 32
3. The Science of God — 64
4. The Social Coefficient of Knowledge — 98
5. The Stratification of Truth — 131
6. The Trinitarian Structure of Theology — 160

GENERAL FOREWORD

A VAST shift in the perspective of human knowledge is taking place, as a unified view of the one created world presses for realisation in our understanding. The destructive dualisms and abstractions which have disintegrated form and fragmented culture are being replaced by unitary approaches to reality in which thought and experience are wedded together in every field of scientific inquiry and in every area of human life and culture. There now opens up a dynamic, open-structured universe, in which the human spirit is being liberated from its captivity in closed deterministic systems of cause and effect, and a correspondingly free and open-structured society is struggling to emerge.

The universe that is steadily being disclosed to our various sciences is found to be characterised throughout time and space by an ascending gradient of meaning in richer and higher forms of order. Instead of levels of existence and reality being explained reductionistically from below in materialistic and mechanistic terms, the lower levels are found to be explained in terms of higher, invisible, intangible levels of reality. In this perspective the divisive splits become healed, constructive syntheses emerge, being and doing become conjoined, an integration of form takes place in the sciences and the arts, the natural and the spiritual dimensions overlap, while knowledge of God and of his creation go hand in hand and bear constructively on one another.

We must now reckon with a revolutionary change in the generation of fundamental ideas. Today it is no longer philosophy but the physical and natural sciences which set the pace in human culture through their astonishing revelation of the rational structures that pervade and underly all created reality. At the same time, as our science presses its inquiries to the very boundaries of being, in macrophysical and microphysical dimensions alike, there is being brought to light a hidden traffic between theological and scientific ideas of the most far-reaching significance for both theology and science. It is in that situation where theology and science are found to have deep mutual relations, and increasingly cry out for each other, that our authors have been work.

The different volumes in this series are intended to be geared into this fundamental change in the foundations of knowledge. They do not present 'hack' accounts of scientific trends or theological fashions, but are intended to offer inter-disciplinary and creative interpretations which will themselves share in and carry forward the new synthesis transcending the gulf in popular understanding between faith and reason, religion and life, theology and science. Of special concern is the mutual modification and cross-fertilisation between natural and theological science, and the creative integration of all human thought and culture within the universe of space and time.

What is ultimately envisaged is a reconstruction of the very foundations of modern thought and culture, similar to that which took place in the early centuries of the Christian era, when the unitary outlook of Judaeo-Christian thought transformed that of the ancient world, and made possible the eventual rise of modern empirico-theoretic science. The various books in this series are written by scientists and by theologians, and by some who are both scientists and theologians. While they differ in training, outlook, religious persuasion, and nationality, they are all passionately committed to the struggle for a unified understanding of the one created universe and the healing of our split culture. Many difficult questions are explored and discussed, and the ground needs to be cleared of often deep-rooted misconceptions, but the results are designed to be presented without technical detail or complex argumentation, so that they can have their full measure of impact upon the contemporary world.

PREFACE

THE chapters that make up this book represent a revised form of *The Harris Lectures* delivered during November and December 1970 in The University of Dundee under the title "God and the World". They were subsequently given again in an adapted form to postgraduate students in Edinburgh under the title "Christian Theology and Cosmological Change". Now that they are at last prepared for publication another title seemed to be required. The title I have now given to the book is meant to bring it into association with other works of mine such as *Theological Science, Christian Theology and Scientific Culture,* and *Divine and Contingent Order.* In spite of being rewritten, however, the material now presented is substantially the same as that in the original lectures. Some references to publications that have appeared since 1970 have been included, mostly in the notes to each chapter, but I have not altered the discussion apart from filling it out where it seemed to be needed today. The title of the last chapter has been changed, as I had already used it in The Richard Lectures for 1978–79 delivered in the University of Virginia in Charlottesville, and published by the University of Virginia Press in 1980 as *The Ground and Grammar of Theology.*

The book is activated throughout by the conviction that knowledge of God is the basic act of the human mind and that faith in its intellectual aspect is the adaptation of the reason in its response to the compelling claims of God as he makes himself known to us in his Word. Spinoza once asserted that once a thing is understood it goes on manifesting itself in the power of its own truth without having to provide further proof. If that is the case, as I believe it to be, then once divine revelation has seized our minds, our understanding of God is carried forward by the intrinsic power of his Truth as it continually presents itself to our minds and presses for fuller realisation within them. Understood in this light, theology is not just a second-order activity of reflection, but a first-order activity of inquiry pursued in a deepening empirical as well as a theoretical relation to the living God. It is a form of intense intellectual communion with God in which our minds are taken captive by his Love and we come to know God more and more through himself. Even though we are found using third-personal language, theological inquiry of this kind is carried out face to face with God so that it may properly be regarded as a form of rational worship in which awe and wonder and joy give vent to themselves in prayer and praise.

This way of regarding theology is not meant to disclaim the place and significance of second-order activity. Constant arduous reflection is needed not only on the content of God's articulate self-revelation as it assumes doctrinal form in our understanding, but on the conceptual reconstruction and adaptation of our modes of thought and speech that must take place, if appropriate conceptual structures are to be developed both in order to help us give coherent and consistent formulation of what we have come to understand and to provide us with fresh intellectual instruments which under the control of the realities we apprehend may serve their disclosure to our continuing inquiry. In this way second-order theological activity is not detached from first-order activity as some sort of uncommitted impersonal reflection upon it, but is pursued only as it is geared into and shaped through first-order activity in direct engagement with God's self-communication to us. These two forms of theological activity will be found interwoven throughout this book, although the primary interest lies with the philosophy of theology regarded as a positive, and not merely a formal or critical, science. The discussion, therefore, will be much more about epistemological structure than with material or doctrinal content. While form and content, method and subject-matter may not be separated, attention will be focussed upon the intellectual aspect of theological science in order to bring to consideration the conceptual inter-face between our knowledge of God derived through his self-revelation and knowledge of the created universe gained as its inherent rational order becomes disclosed through our natural scientific inquiries. That is to say, it is with the philosophy of theological science that we will be concerned, within the changed perspective brought about by the general reconstruction of the foundations of knowledge which we owe to a profound integration of empirical and theoretical factors in recent scientific advances.

It may help the reader if at this point some guide is offered to the contents of the various chapters that follow.

In the first chapter an attempt is made to clarify the difference between the basic attitudes of mind relating to a classical or objective approach to knowledge and a modern or constructivist approach to knowledge. While the former has always characterised realist scientific inquiry, the latter has attained its most widespread expression in the technological society, and in the instrumentalist science with which it operates. Theology and every scientific pursuit operate with the correlation of the intelligible and the intelligent, but how far are we to take seriously the inherent intelligibility of the physical universe of space and time for theology? Christian belief in the dynamic interaction between God and the world, not only in creating but in continuously sustaining its order, demands that full consideration be given to the connection between the rational structures of the created universe and their source in the transcendent Rationality of God. This raises the problem of radical dualism of an epistemological kind which in different ways has long troubled science and theology alike.

The second chapter is devoted to the status of what is called natural theology within this new perspective. Historically natural theology has always come to the front in periods dominated by a cosmological as well as an epistemological dualism, especially evident in mediaeval and post-Newtonian thought, when some sort of rational or logical bridge between God and the world was demanded. While the logical and abstractive procedures which in different ways were employed by mediaeval and modern natural theology are found wanting both on scientific and on logical grounds, the demand for a 'natural' relation between knowledge of God and the intelligibility of the created universe must be met. A move in this direction now appears possible in view of the collapse of abstractive and positivist science before the new realist science that has emerged through the interrelation of four-dimensional geometries and relativity theory. A new kind of natural theology can now emerge, not as an independent antecedent conceptual system, but one which is integrated with positive or revealed theology in the inter-face between Christian theology and natural science.

In the third chapter closer consideration is given to theology as the science of God, in the light of the switch that has taken place in the fundamental conception of science in recent times, in which a closed mechanistic and deterministic conception of the universe has yielded to a dynamic and open-structured understanding of the universe more congenial to the Christian understanding of God in his dynamic and providential relation to the world of space and time. The claim is put forward that theology is a pure science of a realist kind operating on its own proper ground and governed by its own proper object, and comparisons are drawn between theological science and natural science in these respects. Thus understood theology is a positive and progressive inquiry into the knowledge of God proceeding under the determination of his self-revelation but within the limits of our creaturely rationality. It is a human enterprise working with revisable formulations in a manner not unlike that of an axiomatic science operating with fluid axioms.

The social coefficient of our knowledge of God is examined in the fourth chapter, as the implications of the mutual relation between God and man, and of the community of reciprocity in which knowledge of God arises, are drawn out. The relation of all scientific pursuits to the community structures and paradigms in which we think and express ourselves is then discussed, with a view to clarifying more closely the relation of scientific theology to the group habits of thought embodied in culture, society and the Church. Questions are raised about the basic modes of rationality in which our convictions arise and the way in which the social coefficient of knowledge helps to shape the heuristic instruments with which we operate in inquiry. The crucial importance of mystical theology in marginal control of dogmatic formalisations is stressed, and the interrelation of axiomatic procedures to doxological activity in the people of God likewise.

The fifth chapter is devoted to the stratification of scientific knowledge and the hierarchy of truths that arise within it. Attention is given to the multilevelled structure that arises within each science and within the coordination of sciences to one another within the unitary frame of rationality immanent in the universe. In the light of what is learned in this respect from Einstein and Polanyi, but also from Anselm, it is shown that scientific theology operates with different levels of rational complexity, but also with a small core of fundamental concepts and relations, as well as with sets of derived notions and theoretic constructions which have to be cut away when they have served their purpose in the search for economic simplicity in our knowledge of God. The need for a radical simplification and unification of the whole body of received theological knowledge is emphasised.

The final chapter offers a discussion of the trinitarian structure that arises under the constraint of God's self-manifestation to us in history as Father, Son and Holy Spirit. The argument is developed through a discussion of the thought of Augustine, Richard of St. Victor and Thomas Aquinas, with a rejection of the damaging disjunction between knowledge of the One God and knowledge of the Trine God inherited from St. Thomas, and with an acceptance of the idea that it is in the personal and inter-personal character of our knowledge of God that its trinitarian structure arises. The Christian understanding of the person in relation to the personalising activity of the

Holy Trinity is then developed, and an account is offered of its relevance for the openness of our thought to the inherent intelligibility of the universe and for a transcending of the critical splits within the personal and social existence of modern life.

It is many years since Principal James Drever extended to me the invitation of the Court of the University of Dundee to give the Harris lectures. I wish to express my thanks to them for the honour they did me, and to say how sorry I am that it has taken me so long even after my retiral from Edinburgh University in 1979 to find time to get the lectures ready for the press. I look back with great appreciation at the warm and courteous reception which Dundee University gave to a theologian in their midst and the generous hospitality they extended to my wife and myself.

Thomas F. Torrance

Edinburgh,
May, 1982

Chapter 1

CLASSICAL AND MODERN ATTITUDES OF MIND

IN his autobiography Bertrand Russell has told us that in his own view of the world he reversed the process which had been common in philosophy since Kant. "It has been common among philosophers to begin with how we know and proceed afterwards to what we know. I think this is a mistake, because knowing how we know is one small department of knowing what we know. I think it is a mistake for another reason: it tends to give knowing a cosmic importance which it by no means deserves, and thus prepares the philosophical student for the belief that mind has some kind of supremacy over the non-mental universe, or even that the non-mental universe is nothing but a nightmare dreamt by the mind in its unphilosophical moments."[1]

There we have pin-pointed for us one of the crucial issues in modern thought to which we must give considerable attention at the outset of our discussion, for so much depends upon it. What is our basic attitude of mind to the universe around us? How is our knowing related to what we know? It will be the argument of this chapter, and of much that follows, that we must agree with Russell, for while what we know and how we know, subject-matter and method, cannot be finally separated from one another, no true knowledge can be explained by beginning from the knower himself. We do not really know anything unless we can distinguish what we know from our knowing of it; nor do we properly understand what knowledge is about unless we discern in some measure how our knowing is determined by the nature of what we know, as well as conditioned by the activity of the knowing subject. On the other hand, it is also evident that we cannot think or speak of what we know cut off from our knowing of it. In some sense, therefore, our knowing of a thing constitutes an ingredient in our knowledge of it, or at least in the articulation of our knowledge of it. The recognition of this fact can have the salutary effect of preventing us from making inordinate claims about the objectivity of our knowledge, but it also helps to remind us that what we know has a reality apart from our knowing of it. Hence, as Einstein used to insist, "the belief in an external world independent of the perceiving subject is the basis of all natural science".[2]

The relevance of this for our concern, the knowledge of God in his relation to the world, is evident, for in it we have to do with knowledge of God and the world by man who is himself a constituent of the world. By the world is meant not only all that is not God but the unity and totality of created existence, non-mental and mental, non-human and human. It is the universe of space and time both as known by man and as stretching out indefinitely beyond his knowing of it. Since the universe includes man, it includes his knowing of it within the full process of its reality; it is the cosmos of created being in which the relation between knowing and being falls within being.[3] Thus the knowing of being is to be acknowledged as an operation of being itself, for it is through being known that the structure of the universe manifests itself. It is of course to man that the universe becomes known and since he is a constituent of the universe it is in and through him, as Karl Barth has put it, that the universe in this way knows its own being.[4] Thus it fulfils its reality in unfolding its nature and order to our rational understanding. This is the universe with which we have to do in all science, and of which we speak in theology as the creation of things visible and invisible, but of which we cannot speak in science or in theology apart from the process of its manifestation to us or apart from the fact that becomes evident to us in that process, that the universe has much to "tell" us of itself that far outruns our capacity to take it in. The more the universe reveals to our questioning the mysteries of its being and the marvellous beauty of its structures, the more we are convinced that in its own nature it is accessible to rational investigation, and indeed that here we have to do with a rationality so profound that it can be grasped by us only in comparatively elementary forms owing to the limits of our human minds. Thus it remains an ineradicable character of the universe that it cannot be wholly penetrated by our science.[5] It may be finite, but so far as human knowledge of it is concerned the universe is unbounded.

What does this mean? Certainly the universe as we know it is one in which being and knowing are mutually related and conditioned, intelligible reality and intelligent inquiry belong together. But the great question still confronts us. Granted that the universe as we know it constitutes an intelligible whole, and granted that the universe exists, as we say, not only *in intellectu* but also *in re*, is the universe comprehensible to us because somehow it is *intrinsically* intelligible, because it is endowed with an immanent rationality quite independent of us which is the ground of its comprehensibility to us, or is the intelligibility with which the universe is clothed in our knowledge of it something *extrinsic* to it, which we construct out of our own mental operations and impose upon the being of the universe? The most persistent answer to that question throughout the centuries has been that which points to "natural" patterns and structures in the universe which are what they are independent of us but with which our mental operations are steadily coordinated. In modern times, however, there has developed a widespread tendency to hold that the intelligibility of the universe does not originally belong to it but derives from the structuring operations of man's consciousness and is shaped by the ends which he has in view. We shall speak of these as the "classical" and the "modern" attitudes of mind, and correspondingly draw a distinction between *inherent rationality* and *technological rationality*.

The classical attitude of mind, in the form in which it has developed in our western culture, owes a great deal to

classical attitude

the reconstruction of Greek thought through patristic theology and philosophy. It has indeed a distinctively Christian foundation, a fact which is often overlooked by those who think they can dismiss conceptions of inherent rationality by putting them down to a recrudescence of Platonic tendencies in mathematics or even in logic. That Christian foundation for the classical attitude of mind is most clearly seen in the work of the great Alexandrian thinkers of the third and fourth centuries, who set out to articulate the Christian doctrines of the creation of the world out of nothing and the incarnation of the divine Logos within the philosophico-scientific culture they had inherited from the centuries immediately preceding them. They had an immense achievement to reckon with, an eclectic philosophy which tried to work out a synthesis of Platonic, Aristotelian and Stoic thought with a special concern for cosmology, epistemology and logic, a very sharp critical philosophy with its roots in the New Academy, and not least the remarkable flowering of natural science in astronomy and in mechanics, not to speak of mathematics, together with a corresponding development in the grasp and handling of heuristic processes of inquiry.[6] The way in which Christian theologians took over and adapted the cognitive tools and methods of Greek science is specially worth studying in the thought of Clement of Alexandria,[7] but it is to Origen that we may turn for our particular purpose, to see how the Greek attitude of mind was reshaped in the Christian understanding of the relation of God to the world.

Origen accepted the principle, advocated for example by the Stoics, that comprehension and limitation go together, for what is not determinate or limited is incomprehensible. That is why, it was held, God may be spoken of as "incomprehensible", for he is immeasurable and far transcends all our thoughts about him: he cannot be "contained" within the grasp of our creaturely concepts.[8] Origen was Platonist enough to hold that the mind does not need a sensible magnitude in order to think,[9] and agreed that the finite mind cannot think of what is without beginning and without end, but he refused to agree that if we can think God he must be finite. What Origen did, however, was to turn the Stoic idea the other way round, by insisting that it is God's act in comprehending and "containing" all things by his power that limits them, giving them beginning and end, and thus structures them and makes them comprehensible for us.[10] In creating the universe out of nothing God created space and time along with it and thus impregnated what he had made with its rational order. He conferred upon it an immanent or intrinsic rationality,[11] while he himself, the transcendent God, and only he, can be fully and freely immanent throughout the universe without being limited or restricted by it.[12] It was upon this foundation, to which of course others contributed, that classical Christian theology began to be built, but while Origen himself took over a Platonic or a Philonic disjunction between the intelligible and sensible worlds, his successors like Athanasius rejected that disjunction between the cosmic realms as well as the deism that inevitably went with it, and set out instead a dynamic interaction between God and the world which gave the universe a cohesion and unity under God.[13]

This attempt by Alexandrian theologians to think through and set on a sound scientific basis the Christian understanding of the relation of God to the world had a far-reaching impact on the foundations of philosophy and science, as one can see, for example, when the relational view of space and time was carried over by John Philoponos in the sixth century from theology into physics.[14] Three points in particular deserve to be noted. (a) The Judaeo-Christian doctrine of the one God, the Creator of all things visible and invisible, overcame Greek Polytheism and pluralism, polymorphism and dualism, and yielded a unitary view of the created universe which provided a basis for one science and one comprehensive scientific way of knowing that answered to the one pervasive rationality of all created being. It is ultimately to this relation of the one God to the world he has made that one must trace the desire of Raymond Lull, Descartes, Leibniz or even Bertrand Russell to develop some *mathesis universalis* applicable in every area of knowledge, although as we now know a complete and consistent formalisation of such a mathesis has proved inherently impossible. (b) The doctrine of the goodness of the creation, deriving from the Old Testament but reinforced by the doctrine of the incarnation of the eternal Logos or Son of God within the creation, established the reality of the empirical, contingent world, and thus destroyed the age-old Hellenistic and Oriental assumption that the real is reached only by transcending the contingent. The recognition that the temporal and sensible universe has an inherent rationality of its own in virtue of its creation by God, one which God himself takes seriously in the way he relates himself to it, made possible the development of positive, empirical science, and indeed a knowledge of the universe grounded in its own inner determinations and relations. (c) The fact that God himself, the transcendent and creative Source of all rationality, conferred rationality upon nature in creating the universe out of nothing, radically altered the concept of intrinsic intelligibility, for it took it out of the ambience of static structures whether of the Platonic or Stoic sort, and destroyed the Aristotelian separation of terrestrial from celestial mechanics. Its immediate result, of course, was the dynamic and relational conception of space and time as the bearers of rational order in the created universe, and the alteration in the understanding of history which, although it took long to develop, is so characteristic of our western culture. It is to patristic thought that we owe the conception of an ontology in which structure and movement, the noetic and the dynamic, are integrated in the real world. In spite of all that happened in the history of human thought since then, it would seem that it is still for this kind of inherent intelligibility that people are striving, whether they are concerned with the discovery of a quantum logic in physics, or an organismic logic in biology, or with the elaboration of a "structualism" that is more widely applicable across different areas of knowledge and behaviour.[15]

Another way of indicating the contribution of early

Christian thought to the foundations of knowledge would be to refer to the profound integration of logos and being which it discerned, in a transcendent way in the living and active God, and in a creaturely and contingent way in created reality — "being" is to be understood here, not in the old static sense, but as including movement, creative activity in God and becoming or motion in the creature. It was because of the separation between logos and being in Greek thought that it became bogged down, and the promise of its science did not finally mature. Nowhere is this more apparent perhaps than in the relentless attacks by Sextus Empiricus upon the "dogmatics" and the "mechanics", that is, upon those scientists and engineers who asked questions with a view to getting positive answers and putting them to actual use. The kind of sceptical questions he advocated, questions that refused to be committed to positive answers, only ministered to the self-stultification of Greek thought. It was precisely when more than a thousand years later Francis Bacon set that kind of question aside altogether, that learning began to advance in a positive way and empirical science as we now know it got off the ground at last. We shall see that the reason for the long delay in putting into effect the insights in the classical basis of Christian thought is to be found in the remergence of the old dichotomies, but meantime it must be pointed out that beginning with the patristic reconstruction of Greek thought the major emphasis in the Christian tradition continued to be laid upon the *truth of being*, that is, upon the intrinsic rationality of reality. It was the development from Augustine to Anselm that was particularly significant in the West, culminating in a conception of truth in structured levels of reality, to which we shall turn in a later chapter, but several important ideas in that development may be noted here.

On the one hand, the conception of inherent intelligibility means that, whether we are concerned with things visible or invisible, knowledge is to be attained only as we are able to penetrate into the inner connections and reasons of things in virtue of which they are what they are, that is, into their ontic structures and necessities. Only as we let our minds fall under the power of those structures and necessities to signify what they are in themselves do we think of them truly, that is, in accordance with what they really are in their natures and must be in our conceiving of them. On the other hand, the inner relation between logos and being, or the concept of the truth of being, does not reduce to a vanishing point the place or function of the human knower, but on the contrary provides the ground upon which the inseparable relation of knower and known in human understanding can be upheld. This was already made clear by St. Augustine.[16] However, as the history of thought has shown again and again in later eras, no sooner has full place been accorded to the agency of the human subject in knowledge than it tends to arrogate to itself far more than its share. Hence after the problems that became so rife in the Middle Ages, John Major could claim that genuine knowledge may be "active" but it is not "factive".[17] Spontaneous and active though our knowledge of created realities is, it falls under the power of their rational structures which go back to the "creative reasons" freely produced by the mind of God.

Into this whole development, however, there was early injected another element which was to have fateful consequences for the whole history of thought in the West. This is the Neoplatonic disjunction between the intelligible world and the sensible world reintroduced into Christian thought by St. Augustine,[18] which far from being vanquished by Aristotelian metaphysics and epistemology after the twelfth century was actually reinforced by the Aristotelian separation between terrestrial and celestial mechanics and the doctrine of the Unmoved Mover. This had the effect of altering the perspective in which Greek patristic thought had set out the relation between God and the world, for by making all thought pivot upon a point of absolute rest in God, it turned it back into a changeless relation between God and nature so that if the intelligibility of the universe was not concieved in terms of Platonic forms it tended to be conceived in terms of timeless essences and static structures.

In its own way, and with a thoroughness and comprehensiveness never achieved before, mediaeval philosophy and theology in the West gave masterful expression to the Christian conviction of the inherent intelligibility of the universe. The whole of mediaeval culture, art and philosophy, as well as theology, developed and flowered on the assumption of an objective order and harmony in things, which made it possible for human knowledge and life on earth to reach a stability, regularity and beauty reflecting in some measure at least the eternal patterns in the heavens. The magnificent edifice of thought that arose in this way so deepened and grounded the classical attitude of mind as to the rational connections and structures immanent in being, that, even though many of the buildings in the edifice had later to be demolished and reconstructed, the contribution of mediaeval thought to the rational foundations of science endured. Yet at the same time Latin mediaeval philosophy produced many of the problems with which we are still faced in the West. Realist though its foundations were, there took place a progressive shift from the truth of being to cognitive truth, together with an increasing emphasis upon the agent intellect and the rationalism to which it gave rise. Aristotelian notions were at the root of this: that is, not only Aristotle's own shift from the truth of being to a conception of truth located in the intellect,[19] but his insistence that all knowledge is reached only by way of abstraction from sense-experience for there is nothing in the mind which was not first in the senses. This gave rise to the insidious notion of "images in the middle" (a mediaeval form of the doctrine of representative perception), in which the impressions received by the possible or passible intellect are worked up by the agent intellect into "objects of the mind". Thus a radical dualism was introduced into the heart of knowledge, which had the effect of forcing a transition from the truth of being to intellectual truth.

If the immediate objects of our apprehension are not things in themselves but worked up representations or images of them derived from the senses, then it is a correspondence view of truth that seems to become necessary. But on the other hand if we do not apprehend things in themselves, but only sensible images of them, how do we know that they actually do correspond to the objects in our mind? Some method of verification beyond correspondence becomes necessary, and so, in the lack of any transcendental reduction of the conditions under which objective knowledge can be established, there can only take place a retreat into a coherence view of truth. Such a process cannot be halted there, however, as the development of the *via moderna* with its "new logic" made clear, for the rigorous control of propositional statements through stringent formalisation forced a further retreat into a logical and then into a merely linguistic view of truth. And so mediaeval thought was forced at the end of the day into the dilemma between terminism and psychological moralism, for once its dualism (such as that between primary intention and secondary intention or supposition) had broken the referential and ontological relations of concepts and statements, mediaeval thought was compelled to interpret them either in terms of their morphological and syntactic functions in connected discourse or in terms of the subjective states which they expressed in the thinker or speaker: but in both cases oblique intention was substituted for direct intention, and the ground was cut away from all direct and empirical knowledge. In spite of this movement, however, from realism to nominalism and/or moralism, the retreat from the truth of being was held in check by the basic conviction of the immanent rationality of the created universe: dualism, cosmological and epistemological, was not overcome but it was held in check.

The basic ambiguity that this position involves is even apparent in the thought of St. Thomas Aquinas. When he spoke of truth as congruence of understanding and reality (*adaequatio intellectus et rei*), it is left uncertain as to whether it means that understanding conforms to things or that things conform to the understanding. St. Thomas' realist intention certainly demands the former, but there is another side to his thought. In the *Contra Gentes*, for example, he can argue that the act of knowing does not take place by way of encounter with external reality (*per contactum intellectus ad rem*).²⁰ What he seems to have in mind is a basic correlation between being and knowing which allows him to hold, on the one hand, that whatever can be, can be understood (*quidquid enim esse potest, intelligi potest*), that is, to hold to the truth of being (*omne ens est verum*), but it also allows him to hold, on the other hand, that right from the start being is the same as being known, and so he can move over from the concept of the inherent intelligibility of being to the understanding itself as the locus of truth, i.e. from *ens intelligibile* to *intellectus*.²¹ Much the same teaching is to be found in St. Thomas' commentary on Aristotle's *Metaphysics*, in which he can speak of the intelligible and the intelligent as operationally one and as belonging to the same genus.²² This looks like a transition from a realist to an idealist position, but it is held in check by the mutuality between being and knowing which does not allow knowing to take off from the inherent intelligibility of being into a masterful realm of its own, and finally held in check by the doctrine of creation in which God is thought of as providing in himself the ultimate and intelligible ground of all being and knowing. In view of this delicate balance between the intelligible and the intelligent, it is not difficult to discern why Thomist realism was so sensitive to the attack of Occamist criticism, for as soon as it opened up the latent dualism in mediaeval thought, faith and reason appeared ready to fall apart. Once secularised, however, the mediaeval epistemology that depended so much on the elaborations of the agent intellect could only too easily pass over into the kind of phenomenalism that later stemmed from the "Copernican revolution" inaugurated by Kant.

Some of these problems were already discerned by Duns Scotus who reacted rather critically to the teaching of St. Thomas and sought to establish a realist epistemology in which room was made for direct relation with being and for intuitive apprehension of reality, together with a severe curtailment of abstractive processes of thought. Not images in the middle but being as such (*ens inquantum ens*) was recognised as the proper and adequate object of knowledge.[23] It was from Duns Scotus that Martin Heidegger seems first to have taken his cue in tracing the persistent difficulties of Western philosophy to the parting of thought from reality and the domination of abstract formalisation over nature.[24] Already in ancient Greece mind and being, *logos* and *physis* or *ousia*, were moving apart, but it was owing to the extension of abstractive processes of thought that there took place such a secession of reason from being that it set itself up as a law over being and operated by imposing abstract patterns of thought prescriptively upon being. It was this process that was halted and reversed in the early centuries of the Christian era when patristic thinkers laid the basis for the classical attitude of mind, but a similar movement of reorientation began in the fifteenth and sixteenth centuries when Renaissance and Reformation thinkers reconstituted the basis of thought in such a way that modern empirical science was launched upon history. That great shift in epistemic attitude and operation is so well known that we need only glance at certain aspects of it.[25]

1. The renewed emphasis upon creation out of nothing, together with the rejection of the deistic detachment of God from the world, latent in the Latin concept of God's immutability and impassibility, played a primary role, but it must be taken along with the overcoming of the Aristotelian separation between celestial and terrestrial movement which came with the progress of astronomy in the seventeenth century. Together this unification had the effect of replacing the notion of static structures with rational patterns in nature in which the ontic and dynamic aspects of reality were merged. With the basic change in the doctrine of God that stemmed from the Reformation the fundamental concept of nature changed as well: an empirical approach was demanded, and objectivity was respected.

2. At the same time a new kind of question arose,

replacing the problematic *quaestio* which had become the regular instrument of scholastic theology and science in the solution of epistemological difficulties and in the clarification of knowledge. The latter was the kind of question which could only be answered with formal logic, the so-called *ars diiudicandi*. But now a genuinely interrogative question came to the front, in which inquiry was initiated in order to disclose facts and truths which could not be logically inferred from what is already known or be established merely by eliminating what is false by means of the principle of non-contradiction. This was the kind of question that could be answered only by some "logic of discovery", the so-called *ars inveniendi*.[26] As John Calvin operated the new questioning it meant deposing the question as to the abstract essences or quiddities of things (*quid sit*) from its place of primacy in scientific procedure, and putting in its place the question as to the actual nature of what is being investigated (*quale sit*), and then the critical question always necessary in scientific knowledge (*an sit*) became no longer a question as to abstract possibility of things but a question testing the grounds of actual knowledge.[27]

3. The effect of this new approach was to throw inquiry directly upon being itself, so that being in its objective reality and self-evidence was given priority over all precedent knowledge or opinion about it. Thus there became entrenched a way of knowing in which people were determined to think as the facts compelled them to think, or to think strictly in accordance with the nature and activity of the given reality. This was not arbitrary, speculative thinking, but thinking bound to its chosen field, but for that very reason thinking that detaches itself from unwarranted preconceptions and prejudices and operates on its own free ground, thinking that acknowledges only the authority of its object and will not submit to any kind of external authority. This is rigorously objective thinking which will take nothing for granted in its determination to be real but which operates by allowing the object being investigated to disclose itself in its own state

and light, by penetrating into its inner connections, and by establishing knowledge in terms of principles and laws which it derives from those inner connections. Thus there grew up a form of science in which reality itself in the last resort was respected as the supreme judge of what is right and true. This is an objectivity that is grounded in the depth of being itself, and not a false objectivism in which the reason is detached from what it seeks to know only to be incarcerated in static formalisms of its own constructing. That is to say, here we have a powerful rehabilitation of the truth of being, together with the classical attitude of mind directed to the intrinsic intelligibility of things, but now cast into an operational form in accordance with the actual nature of the created universe as it becomes progressively disclosed to our questioning.

4. Scientific knowledge of this kind respects the agency of the human knower, and makes room for him without detracting from the principle of objectivity. Far from being impersonal, objective knowledge requires the interaction of persons with the world in order to reach an understanding and interpretation of nature in its dynamic, self-ruling structures, for it is the person, as we shall see later, who is the agent of objective operations. It is not surprising that this insight arose out of theology, for in it, as Calvin pointed out, the object is God speaking personally (*deus loquens in persona*), and knowledge of him arises only as mutual relations are set up between God and the human subject.[28] The fact that the object addresses man means that it calls him into a personal relation with it, but the fact that it is God himself who thus addresses man through his Word means that man is summoned to learn of God out of God's disclosure of himself, and not out of the states and conditions of man's self-knowledge. At the same time it becomes evident that man does not really know God until that knowledge strikes so deeply into man's own personal being that he comes to a new and truer knowledge of himself. In the Reformation itself this demanded the rejection of the idea that the criterion of truth is lodged in the subject of the knower or the interpreter. Hence in the

interpretation of the Scriptures we are thrown back upon the truth of the Word of God which we must allow to retain its own weight and majesty over against us and which we may understand only as we allow the Word to declare itself to us and thereby also call in question all our preconceptions and vaunted authorities. It was for this reason that Calvin could insist that authentic knowledge arises only through a relation of fidelity (*fides*) to what we seek to know, and therefore proceeds by way of voluntary obedience in which we serve it. This is precisely the point that was transferred by Francis Bacon to natural science in his notion of the scientist as a "servant of nature", following the clues that nature provides of its own inner connections and mysteries. The kingdom of nature or the dominion of science is not entered by dictating to nature but by following its own leading: but all this means that a movement of the will has entered into the basic activity of science. Active knowledge and willing obedience to the claims of what we seek to know belong together.[29]

Such, then, was the reconstituted foundation for the classical attitude of mind with which the new science set out upon its centuries of advance. Yet it was not without seeds of trouble from the start, for modern thought moved forward with latent ambiguities similar to those that lay embedded in mediaeval thought. The appeal of the Reformers to the teaching of St. Augustine, the recognised *magister theologiae* of the Western tradition, opened up the way for a revival of the old distinction between the intelligible and the sensible worlds. Indeed Protestant philosophy could well be expounded as a secularised form of Augustinianism. Moreover the Renaissance thinkers reintroduced the old Greek idea of the autonomy of the reason together with Neoplatonic metaphysics. The effect of this was to bring in an ominous tension between reason and experience, and so before very long in different but complementary ways Cartesian and Newtonian thought developed a radical dualism and built it into the basic orientation of Western thought: a dualism between subject and object on the one hand, and a dualism between God and the world on the other hand.³⁰ The combination of these was particularly damaging, for the effect of the detachment between subject and object within a cosmology in which space and time as the divine container were given an absolute status independent of material existence but causally conditioning it as an inertial system, was to reduce the knowing subject to inner states of consciousness over against a determinate nature as the object. Once again something like "images in the middle" were forced forward and a powerful doctrine of "representative perception" took the field, and there followed a retreat from the truth of being into the mental processes and consciousness of the self. And so we come to the valiant Kantian attempt to establish the conditions of epistemic objectivity through transcendental reduction, and to the methodological fiction of "sense data" devised by the empiricists. Thus when after Newton there developed a notion of science as abstraction from observations and it became evident, as with David Hume, that no explanation or demonstration of causality could be reached this way, there took place the "Copernican revolution", as Kant called it, to account for Newton's laws of motion in terms of the structures and operations of the human mind. This was tantamount to transferring Newton's inertial system to the human mind. Space and time were regarded as *a priori* forms of intuition independent of experience, and were thus beyond any possibility of change or modification by experience, while nevertheless conditioning what we experience. They imported a rigidity into the structure of the scientific consciousness, and thereby lodged in it hidden idealist presuppositions which, as history has shown, have been extremely difficult to dislodge especially when combined with a positivist, conventionalist concept of science, which was what actually, and perhaps inevitably, happened.³¹

Two fateful tendencies are then found to be at work in European thought. (a) The transfer of intelligibility to the human pole of the knowing relation, when the concept of the inherent intelligibility of the universe began to fade away. As Kant himself taught, "The understanding does not derive its laws (*a priori*) from, but writes them into, nature".[32] (b) The rise of the masterful idea that we can understand and verify only what we can make and shape for ourselves. To refer to Kant again, he argued that what is unknowable cannot be constructed, and what is constructible is knowable, for it became evident to him in his attempt to make the laws of motion intelligible to the reason that construction is a basic ingredient in defining what is known and knowable.[33] As we look back over the history of this period of European thought, however, it appears evident that abstraction from observations leads sooner or later to constructions out of our consciousness, and then to impositions of our formal patterns of thought upon nature, and that from this dominance of the mental over the non-mental world there arises the notion of instrumentalist science with its powerful tool the technological rationality.

This is the "modern" attitude of mind, the mentality of *homo faber* so characteristic of our industrialist and technological societies, which has penetrated into vast areas of our culture. A curious but instructive instance of this attitude of mind is to be found in Giambattista Vico's account of historical method. He held that the universe is inherently intelligible, but properly knowable only to God who created it. Geometry is rationally accessible to us, for it is our human construction, and physics would be equally accessible to us, if we were to make physical realities, for then we would be able to get inside them and understand their elements and causes: but that is what we cannot really do.[34] What we cannot construct we can neither know nor demonstrate in any proper sense. And so Vico enunciated his celebrated principle of truth: *verum et factum*.[35] "The rule or criterion of truth is to have made it."[36] Quite consistently, therefore, Vico turned his attention away from the world of abstractions (i.e. the Cartesian abstraction of form from the nature of objects, of geometry from matter) into which, he claimed, contemporary science was being diverted, to the actual world which men themselves *have made*, the world of nations and human affairs, including their religion and language, civil, moral and political institutions, the world in which men have in a certain sense "created themselves" and within which they continue to be preserved.[37] Since that is our human world, "its principles are to be found within the modifications of our own human mind".[38] That is to say, this world constitutes the kind of reality with which we have an inward affinity by reason of our own experience, feeling and imagination, and into which our *understanding* can penetrate, enabling us to determine its inner causes, connections and forms, and so to formulate, as science requires, universal and eternal principles.[39] It was with this in view that Vico projected his "new science concerning the common nature of the nations", while the master key to this new science he claimed to find in the activity of the "poets" or "creators" who are found in the birth of nations, revealing the poetic wisdom or the creative mind that has shaped the ground and pattern of human institutions throughout the ages. The appropriate instrument for this new science is not the analytical but the synthetic use of reason, not the critical method but *invention*.[40]

Vico is significant for he presents us, if in a somewhat exaggerated form, the characteristic attitude of mind that was to infect, not only the creation and appreciation of works of art, but the literary and human sciences as well. But the will to acknowledge and accept as valid only what we can construct and control has made deep inroads into the physical sciences also, although there it comes up against areas of human inquiry into the states of the universe where no construction is possible, as in astrophysics, and other areas where the more man tries to control nature the more he seems to get in his own way, as in microphysics where it is claimed that the scientist cannot get behind his observational interference with nature. At these points the hidden idealist presuppositions in the roots of abstractive science come to the surface, as in the idea that the mathematics with which we bring to orderly representation determinate realities is a pure construction of the human mind, or in the conviction that there is no final way through from the observing subject to the object, so that some way of transcending the subject-object relation is demanded. Thus there sets in a reaction to direct intention in the pursuit of knowledge and a retreat into oblique meaning such as we find in existentialism.

It is now apparent that the programme *de omnibus dubitandum est*, with which modern thought set out so bravely from Descartes, only results in the alienation of man from the world. When thought and reality part company, the world becomes opaque and meaningless no matter how coherent the clear and distinct ideas we have generated, for the more man tries to force clear-cut meaning of his own devising upon the world, the more he cuts himself off from nature and is apt to misuse it for his own ends. But the more he tyrannises over nature and enslaves it to his own ends like that, the more he creates ecological chaos and at the same time finds himself imprisoned in his own constructs, whether they are physical or social mechanisms, and then frustration, nihilism, iconoclasm, and eruptions of violence are more or less inevitable, for man has reduced himself to a thing. The widespread loss of meaning evident here is surely due to man's estrangement from reality in the chasm that has opened up between himself and the universe, but also due to the fact that he has attempted to create meaning for himself instead of discovering it. Patterns may be created but not meaning, for meaning appears only when we attend away from ourselves to the objective ground of our being, and to the source of its intelligibility. Thus the loss of meaning and loss of belief in God appear to go together, the one affecting the other, for the deistic disjunction between God and the world and the alienation of man from the world both throw man back upon himself where he is shut up within his own mental constructs, and upon the boredom and futility that involves. From beginning to end it would appear to be the story of the active reason overreaching itself in a world that is "man-made", but

20   REALITY AND SCIENTIFIC THEOLOGY

which as such is without signification beyond itself and therefore finally empty in itself.

We have been thinking of this attitude of mind, in which man creates meaning for himself in the world by imposing thought-forms of his own construction upon nature, as "modern" in contrast to the classical attitude of mind, but how far are we justified in speaking of it as distinctively characteristic of modern thought? It would certainly appear to be the attitude of mind that permeates much of our modern culture, and not least the social sciences, but this can hardly be said of the pure or exact sciences which in their development have had to fight their way out of abstractive procedures and instrumentalist conceptions of what science is about, and operate on the assumption that the universe is somehow comprehensible in itself and that as such it is the source of the concepts we use and the determining force in the way in which we build them up into scientific structures appropriate to the nature of the universe. That is to say, they are ultimately concerned with what Max Planck and Albert Einstein from different approaches called "reality",[41] with the universe as it really is independent of the observer, an "external world" that is inherently rational, endowed with an intrinsically intelligible nature that makes it accessible to scientific investigation and interpretation. In other words, modern exact science still operates on the basis of the classical attitude of mind which developed through Christian theology and philosophy. It is not surprising, therefore, that Max Planck could end his essay on "Religion and Natural Science" with the following statement. "Religion and natural science are fighting a joint battle in an incessant, never relaxing crusade against scepticism and against dogmatism, against disbelief and against superstition, and the rallying cry in this crusade has always been, and always will be: 'On to God!' "[42]

There was, however, a distinct difference between Planck and Einstein in their attitude to the real world. Max Planck still operated with idealist presuppositions of a Kantian sort, and was aware of them,[43] but this meant for him that though the physicist intended reality in the full sense he could not fully break through to it, and so was forced to operate with a concept of reality that was rather "metaphysical". Planck's own operational principle was that what is measurable is real. In view of the fact that he developed the first exact determination of the absolute magnitude of atoms, and showed that the structure of nature was governed by the universal constant h, he was not without considerable success in penetrating through the phenomenal to the real. The effect of his work, as Einstein has pointed out, was to shatter the whole framework of classical mechanics and electrodynamics and to set science the fresh task of finding a new conceptual basis for all physics.[44]

The really decisive advance, however, was due to Einstein himself, in the establishing of mathematical invariances in nature irrespective of any and every observer, in which he was able to grasp reality in its depth. This was decisive not only because it broke through the idealist presuppositions stemming from Kant but because it broke through the positivist concept of science. We shall have to devote fuller discussion to the problem of positivism later. At this juncture we must content ourselves with noting that while positivism involves (a) a mode of knowledge by way of abstraction from observations, with its implication of a logical bridge between our concepts and observational experience, and (b) a view of scientific theory as a convenient functional arrangement between our observational experiences and the concepts we abstract from them, both these aspects of positivist science have been overturned by Einstein, not only in his theoretico-scientific writings but in his actual achievements and in the way he carried them out. He showed that while we do and must operate with a correlation between the theoretical and empirical components of science, we cannot derive the former from the latter by logico-abstractive processes. As F. S. C. Northrop has written: "This means that the positivistic theory that all theoretic meanings derive from empirical meanings is invalid.

Furthermore, the thesis that the theoretically designated knowledge gives us knowledge of the subject matter of science and of *reality*, rather than merely knowledge of a subjective construct projected by a neo-Kantian kind of knower, confirms the thesis that the thing in itself can be scientifically known and handled by scientific method. Thus *ontology* is again restored, as well as epistemology, to a genuine scientific and philosophical status."[45]

For some time now we have been aware of a serious state of affairs in our Western culture, for a deep fissure has opened up within it. This is not the division between "the two cultures" which C. P. Snow has brought to prominent notice, that between a literary culture and a scientific culture, although it certainly has something to do with that, for it is not a split brought about by educational habits or that has arisen on social grounds, but one that has to do with the foundations of knowledge and different habits of mind.[46] It is the split in our Western culture in which Christian theology, at least in its historic development, and pure science stand on one side, and in which the social sciences and technology (as it arises out of an instrumentalist conception of science) stand on the other side. Both sides are affected by the split, and are distorted by detachment from interaction, but it is in the social science/technology side of the split that the rift is most damaging, especially as there is thrown up in it a tragic contraposition between objectivism and existentialism, mechanism and subjectivism, or even in some quarters between "establishment" and "revolution". The profound epistemological split which affects the foundations of our culture opens up secondary issues which point beyond pluralism to widespread fragmentation and disintegration. In recent decades outbursts of violence in different parts of the world between technological societies and forms of nihilism appear to threaten us with an era of new mechanised barbarism. If the split in our culture persists it can only spell the death of our civilisation.

This unhappy state of affairs cries out for incisive and

sustained rethinking of the very basis and structure of our culture, in the hope that it may be set upon adequate scientific foundations in which the numerate and literate aspects of our human life and activity do not bifurcate but retain their necessary and fertile correlation. Self-regulating devices will have to be built into the fabric of our institutions so that in the development and transformation of human knowledge the self-centredness and self-assertion of human groups and individuals are held in rational check. It will certainly require an immense effort of combined operation among the arts and sciences to prevent the imposition of people's selfish and greedy ways upon nature, if the rapidly developing ecological crisis is to be obviated. But even more a new openness of the human mind to what transcends it, and a new respect for the nature of created things and their openness to human inquiry and interaction are required. What is needed is a recovery of the classical attitude of mind toward the universe, and a modern adaptation of it across the whole spectrum of our culture, which will give rise to a synthesis transcending in its range and effectiveness that which informed the mediaeval world with its meaning. It will have to be a very different kind of synthesis, one that functions through an open hierarchy of different levels or systems in which the manifold fields of knowledge and life can be coordinated without pluralistic fragmentation, and one therefore which will involve a grasp of the comprehensibility and harmony of the universe in a range of depth that cannot be mastered by our formalisations and which will be open enough to foil the imperialism of mechanistic concepts. At the same time it must be one in which the constancy, reliability, integrity and beauty of the universe can be reflected in human civilisation in such a way as to give it unity, stability, continuity and meaning. The world of human achievement requires an Archimedean point beyond it by which it can be steadily levered out of its own self-incarceration, and that means that it must be coordinated with the openness of all created being to the unlimited reality of God.

Is all this possible without a return to a Christian understanding of the relation of God to the world he has made, which gave rise in the first place to the classical attitude of mind which we have been considering? It is the transcendence and oneness of God that gave unity, identity, objectivity and comprehensiveness to the space-time medium in which we and all created reality are bracketed together in one world. It is because all rationality in the universe has a single source in the Creator, and because therefore the different modes of created rationality, number, word, organism and beauty, all interpenetrate one another, that there is a profound unity in all the arts and sciences, but in the nature of the case a unity that will appear properly only in stratified structures of openness to the unlimited reality and uncreated rationality of God.

That is the claim of Christian theology, but it is not without its own serious problems. The doctrines of creation out of nothing and the transcendence of God over all that he has made, mean that the intelligibility which he has conferred upon the universe is not an extension or an emanation of his own but a creaturely intelligibility utterly contingent upon his own, yet somehow coordinated with it as the universal medium through which he may be known. However, the contingent character of this intelligibility and its intrinsic relation to being created out of nothing also mean that we cannot argue from the creaturely intelligibility to the uncreated intelligibility of God any more than we can operate with a logical bridge between the created universe and the Creator. The doctrine of the creation of the world out of nothing implies that if we are to understand the world of contingent reality we must investigate the world itself by itself, and learn of it from out of its own natural processes, that is, by attending to its contingent nature and created rationalities. By doing that, however, we look upon the world, without taking God into account among the data. However, if this is the case, does the world not then lose its meaning? That is a problem we must take up for consideration in the next chapter where we discuss the status of what is called "natural theology" in the world of modern scientific knowledge.

There is also a problem, here, however, for what is called "positive theology" which is after all a human, creaturely undertaking and as such falls far short of the transcendent intelligibility of God *quo maius cogitari nequit*, as St. Anselm reminds us, and which therefore cannot claim that God *must be* the content of its intelligible forms. If our theological concepts do terminate upon God, they do so only in virtue of his grace, and cannot therefore import any kind of necessary or logical relation to him even though they arise in our understanding under the pressure of God's self-disclosure. This is not to say that there is no correlation between the knowability of God and our human knowledge of him and its theological elaboration, but that that correlation, and indeed the mutual relation between the intelligible reality and intelligent inquiry which it involves, repose upon the free ground of God's own Being. The importance of this for theological science can hardly be overestimated, for it is through such a relation between its creaturely intelligibility and the transcendent intelligibility of God that theology may be freed from being trapped within its own formalisations and their inevitably time-conditioned character.

This problem may be illustrated by reference to Paul Tillich's distinction between what he called "subjective reason" and "objective reason".[47] Subjective reason enables the mind to grasp and shape reality, and objective reason is the rational structure of reality which can be grasped and shaped. Here the circle of knowing starts from the human knower and takes in being and then returns to the knower, for what is known is not only grasped but shaped by him. Tillich was aware that a proper circle of knowing can easily become a vicious circle, and sought to guard it through other notions such as the "ecstatic reason", that is, the reason as it is gripped from beyond or outside of itself. However, this seems finally to be nullified by the principle of correlation as Tillich used it, for from beginning to end it means that God is correlated with the question as to God which man himself is precisely as man. Since this does not seem to involve any questioning of our ultimate questions, there does not seem to be any possibility of breaking free from the closed circle of correlation. That is to say, it is highly questionable whether Tillich ever managed to break through a Kantian or phenomenological notion of being to a genuine ontology in the sense in which it has been established by Einstein in natural science. Once again we are forced to draw a distinction between psychological movement and an epistemological movement from us to reality independent of ourselves. When we come up against something which we cannot shape (such as the invariant structures in the universe which are what they are independent of our conceiving of them), although we may and do shape our understanding and articulation of it, our psychological movement in the process of inquiry is inverted or converted into an epistemological movement. Although we start off from some point where we already are, we come right up against being in such a way that it turns our movement of thought round in a new start from being to us and back again, so that the circle of our knowing does not, so far as its material content is concerned, begin from and rest upon us but upon what is other than us and objective to us. It is this *epistemological inversion* of our psychological procedures to which we pay attention when we pose our scientific questions: they start from where we are and so with some frame of thought which we already occupy, but they are posed as far as possible in an open way, i.e. without being closed from behind by fixed presuppositions inherent in our starting point. Hence as we direct our questions to our chosen field we allow it to disclose itself to our inquiry, and as that takes place we proceed to question our initial questions, and then we pose our revised questions to the field and in the light of what further becomes disclosed we requestion our prior questioning, and so on. Thus scientific inquiry operates in such a way that it cuts back constantly into ourselves the questioners, in order to invert the determining factor from ourselves to what we seek to know. This is why rigorous scientific inquiry far from being some sort of impersonal progression of induction is a highly distinctive movement of interaction of the inquirer with the object, in which acts of personal self-criticism and personal judgment are called for all through the process of distinguishing what we know from ourselves and of checking the illegitimate projection of ourselves, our subjective states and conditions, into what we seek to know.

Now of course we do not proceed in this way unless we could have some initial glimpse, and some initial grasp, however tenuous, of reality, and unless reality were comprehensible in itself apart from our perceiving or knowing of it, that is, unless it had its own intrinsic relations and structures, for it is only as we are able to hook our thought on to those that we can advance in our inquiry or climb up into fuller knowledge of the reality under investigation. In so doing we presume that a correlation is possible between our human conceiving and the inner structure of reality itself, and we carry out all our operations in that belief. However, that very presumption makes us direct our critical questioning back upon ourselves to make sure that we are not moulding reality in terms of our own constructions or imposing artificial structures of our own upon it.

There are questions arising here to which we shall give fuller attention later as the argument proceeds, but by way of concluding this chapter it may be stressed that it is through images and concepts we use, through the forms of signification our propositions entail, and through the structures that arise in our understanding, in their co-ordination with experience, that <u>there *shows through* an objective rationality which is independent of our forms of thought and speech.</u> In so far as we can distinguish it from them, we have a firm base from which to put our forms of thought and speech to the test, to see how far they actually are coordinated with the realities which they claim to indicate and so provide the intelligible medium in which our minds come under the compulsion of those realities and take shape under their ontic necessity and intrinsic intelligibility. Granted, then, that the structures of our understanding, and the forms of thought and speech which we use in knowledge of God, fall far short of his transcendent Reality so that we may never read their creaturely content into the divine Being, nevertheless as forms of created intelligibility deriving from the uncreated intelligibility of God, they are sufficiently relevant while pointing utterly beyond themselves to serve our understanding and speech of God. Formally, at any rate, this coordination of our concepts with experience is not different from that with which we operate in our inquiry into the created universe; the difference arises in respect of the nature of the divine Being who is the *creative* Source of all our concepts of him in a way that creaturely being cannot be.

NOTES

1. B. Russell, *My Philosophical Development*, London, 1959, p. 16.
2. A. Einstein, *The World as I See It*, London, 1953, p. 60.
3. Cf. Julian Hartt, *Being Known and Being Revealed*, Stockton, California, 1957, pp. 24ff.
4. Karl Barth, *Church Dogmatics*, IV.3, Edinburgh, 1961, p. 140.
5. See A. Einstein, *op. cit.*, p. 5.
6. Cf. G. Sarton, *A History of Science*, vol. 2, *Hellenistic Science and Culture in the Last Three Centuries B.C.*, Norton, New York, 1970.
7. See especially *Stromateis*, bk. viii; and also "The Implications of Oikonomia for Knowledge and Speech of God in Early Christian Theology", in *Oikonomia. Heilsgeschichte als Thema der Theologie*, edit. by Felix Christ, Hamburg, 1967, pp. 223–238.
8. Origen, *De Principiis* (GCS edit. by P. Koetschau, Leipzig, 1913, vol. v), 20.5ff; 21.10ff; 54.4ff; 55.1ff; 164.3f; 272.7; 345.23f; 346.11f; 359.16f; 360.1f.
9. *De Princ.*, 21.13f; 22.4ff; 23.1ff.
10. *De Princ.*, 272.16f; 359.16f; 360.1ff.
11. *De Princ.*, 9.13f; 21.13f; 50.14f; 86.5f; 159.4ff; 271.12ff; 273.1ff; 289.11f; 347.15f; 359.9f.
12. *De Princ.*, 108.11ff; 124.1ff; 140.25ff; 141.1ff; 191.1ff; 283.5f; 351.18f; 352.4ff; 353.5f; *Contra Celsum* (GCS edit., vol. 1), 275.9ff;

277.26ff; 282.18; 365.19; (GCS edit., vol. 2), 184.17; 186.8; 284.14f.
13. Athanasius, *Contra Gentes*, edit. by R. W. Thomson, Oxford, 1971, especially 40–44, pp. 108–122.
14. *Ioannis Philoponi in Aristotelis Physicorum libros commentaria*, edit. by H. Vitelli, Berlin, 1887, pp. 504ff; and the long *corollarium de loco*, pp. 557ff. See further his work *De mundi creatione, Bibliotheca veterum patrum*, edit. by A. Gallandius, Venice, 1778, vol. xii.
15. See Jean Piaget, *Structuralism*, New York, 1970, pp. 4f, 30, 40ff, etc.
16. St. Augustine, *De Trinitate*, 9.12.18: *ab utroque enim notitia paritur, a cognoscente et cognito*. Augustine seems to have applied this even to the concept of number, *De Trin.* 12.2.2.
17. John Major: *notitia activa, non factiva, Comm. in Sent. prol.* q.6, fol. xiv.2, xv.3. See *Archives de Philosophie*, Paris, Avril–Juin, 1970, Tom. 33.2, p. 277.
18. See, for example, *Contra Academicos*, iii.xvii.37; or *De Trinitate*, 11.1.1, or 15.12.21.
19. Cf. the passage of Aristotle's thought from *Metaphysica*, G 7, 1011 b 25 and *De Interpretatione*, 18b, to *Metaphysica*, E 4, 1027 b 25.
20. St. Thomas Aquinas, *Contra Gentes*, II.98.
21. *Contra Gentes*, I.87, ad 3; II.98–99. Cf. also Karl Rahner, *Hearers of the Word*, London, 1969, pp. 41f, 48f.
22. St. Thomas Aquinas, *Comm. in Met., proem.*; II. *Met.* 1.1, n. 280; VII *Met.* 1.1, n. 1304.
23. John Duns Scotus, *Ordinatio*, Civ. Vaticana, 1950, *prol.* n. 1. See further *De doctrina Ioannis Duns Scoti*, vol. iv, "Intuitive and Abstractive Knowledge from Duns Scotus to John Calvin", Rome, 1968, p. 293.
24. See M. Heidegger, *An Introduction to Metaphysics*, Newhaven and London, 1959, pp. 178f.
25. Cf. my earlier discussions, *Theology in Reconstruction*, London, 1965, ch. 4, and *Theological Science*, London, 1969, ch. 2.
26. This distinction derives from Stoic concepts of scientific method, and was reintroduced into European thought by Lorenzo Valla's studies in law under the influence of Cicero. The same influences are discernible at work in the thought of men as different as R. Agricola, Calvin, Copernicus, and Bacon.
27. John Calvin, *Institute*, 1.2.1, etc.
28. John Calvin, *Inst.* 1.1f.
29. Cf. *Theological Science*, pp. 69ff.
30. This double dualism is particularly evident in the thought of John Locke through whom there arose the problems discerned by David Hume. See my discussion of this in *Transformation and Convergence in the Frame of Knowledge*, Ch. 1, "The Making of the 'Modern' Mind from Descartes and Newton to Kant", Belfast, 1984.
31. See especially Ernst Mach, *Die Analyse der Empfindungen und das*

Verhältnis des Physischen zum Psychischen, 1886, and *Erkenntnis und Irrtum*, 1905.

32. I. Kant, *Prolegomena to any Future Metaphysics*, 36.
33. I. Kant, *Critique of Pure Reason*, Bx-XII, B138, A221, B271, A240, B300, A713-4 & B741-55, A734 & B762. Cf. G. Buchdahl, *Metaphysics and the Philosophy of Science*, London, 1969, pp. 555, 572, 626ff.
34. Giambattista Vico, *De nostri studiorum ratione*, 1709, 1709, sect. iv, pp. 84f; tr. by E. Gianturco, *On the Study Methods of our Time*, Indianapolis, 1965; and *De antiquissima Italorum sapientia*, 1710, p. 62.
35. Vico, *De ant. Ital. sap.*, title of first chapter. *Opere*, vol. III, p. 131. See the discussions by Nicola Baldoni, "Ideality and Factuality in Vico's Thought", pp. 391-400, and by Max H. Fisch, "Vico and Pragmatism", pp. 401-424, in *Giambattista Vico. An International Symposium*, edit. by G. Tagliacozzo and H. V. White, Baltimore, 1969.
36. Vico *De ant. Ital. sap.*, p. 62.
37. Vico, *Scienza Nuova*, 1725, tr. by T. G. Bergin and M. H. Fisch, Cornell, 1970, pars. 332, 367.
38. *Ibid.*, par. 331.
39. Isaiah Berlin points out that here we have the source of the German historicist *Verstehen*, historical Einfühlung, etc., *International Symposium*, p. 375.
40. Vico, *On the Study Methods of our Time*, ch. III. Cf. the essay in *An International Symposium* by E. Grassi, "Critical Philosophy or Topical Philosophy?" pp. 39ff, who rightly links Vico's thought to Cicero and Quintilian, but strangely misses the significant contribution of Lorenzo Valla and Rodolph Agricola. The same applies to the essay by G. Cotroneo, "A Renaissance Source of The *Scienza Nuova*: Jean Bodin's *Methodus*", pp. 51ff.
41. Max Planck, *Where is Science Going?* London, 1933, p. 82; *The Universe in the Light of Modern Physics*, London, 1937, pp. 8, 15; *Scientific Autobiography and Other Papers*, New York, 1949, p. 107f; Albert Einstein, *Out of My Later Years*, pp. 58ff. See Henry Margenau on "Einstein's Conception of Reality", in P. A. Schilpp, *Albert Einstein: Philosopher-Scientist*, New York, 1949, pp. 245ff. For the reaction of one still operating under positivist assumptions, see Moritz Schlick, *Space and Time in Contemporary Physics*, London, 1920.
42. Max Planck, *Scientific Autobiography*, p. 187.
43. Max Planck, *Where is Science Going?*
44. A. Einstein, *Out of My Later Years*, "Max Planck in Memoriam", p. 209.
45. F. S. C. Northrop, "Einstein's Conception of Science", in P. A. Schilpp, *op. cit.*, p. 407 (italics mine).

46. This way of thinking of the split culture is to be affirmed in spite of C. P. Snow's argument to the contrary, *The Two Cultures and a Second Look*, 1969 edit., p. 67f. Unfortunately he takes no real account of epistemological divergences.
47. Paul Tillich, *Systematic Theology*, vol. 1, London, 1953, pp. 75ff, cf. p. 53f.

CHAPTER 2

THE STATUS OF NATURAL THEOLOGY

IN the last chapter we considered a contrast between the classical attitude of mind in which we assume, in all branches of objective knowledge, that the universe is possessed of rational structures apart from what we make of it in our scientific operations, and another attitude of mind which has gained considerable hold in the modern world, except in pure science and rigorous theology, in which we assume that it is we ourselves who by our scientific operations clothe the universe around us with form and structure. This is a contrast between finding and creating meaning. An attempt was then made through some analytical probing into the development of Western thought to show that the abandonment of the classical attitude of mind proves disastrous, for it results in a progressive estrangement from nature and a widespread loss of meaning, together with ecological chaos in the natural and social realms of creaturely existence. As soon as we think we can make meaning and impose it at will, we become imprisoned in ourselves and then feel suffocated, which inevitably leads to violent attempts to emancipate ourselves from the tyranny of our own social and technological constructs.

An attempt was also made to show that pure science in our own day has been breaking through the "modern" attitude of mind, to adopt an attitude of mind basically similar to that which is to be traced in the classical foundations of Christian theology and empirical science. This is a movement of thought in which we have restored an objective approach to nature, and an understanding of our scientific enterprise in which we are concerned to grasp reality in its intrinsic rationality. Consequently a call arises for a reconstruction of the whole of our western way of life and thought upon the inherent reasonableness of the universe as it comes to light in its own inner balance and self-regulative processes. In order to achieve this we need to develop a keener intuitive grasp of the comprehensibility and harmony of the created order which allows us to penetrate into its inner connections and patterns deeply enough to grasp something of their natural, objective cohesion, and yet to retain appreciation of the fact that they are not to be exhausted or completely mastered by our theoretical formalisations. We have to be on our guard against the imposition of artificial homogeneities upon the world, a temptation which regularly arises out of the craving of classical mechanics for necessitarian modes of thought. What we need, as Christian theology has always claimed, in laying the foundations of our Western culture and science, is an understanding of created being as correlated to the unlimited Reality and Freedom of God. This universe of ours has been endowed by God, in a created correspondence to his own inexhaustible Rationality, with a subtle and flexible rational order which is always taking us by surprise, as it discloses to our inquiries riches and depths of intelligibility which make us dissatisfied with our scientific formalisations, pointing us far beyond what we already know and beyond what we can ever fully know.

It is at this point, however, that theology finds itself confronted by a fundamental problem. Its concept of the creation of the world out of nothing implies that if we are to understand the world of contingent reality we must investigate the world itself, and learn of it out of its own natural processes and interior relations. By doing that, on the other hand, we have to attend to the world in its created independence, in its own given reasonableness detached from its Author, that is by considering it in its utter otherness from God and without taking God into account.[1] It is the Judaeo-Christian doctrine of God's creation of the world out of nothing that forces this approach upon us, but when we look at the world exclusively in this way it appears to lose its meaning, for then it is not regarded as pointing beyond itself. Here we have a problem basically similar to that which Polanyi has so often pointed out in somewhat different contexts. "So long as you look *at* X, you are *not* attending *from* X to something else, which would be its meaning. In order to attend *from* X to its meaning, you must cease to look *at* X, and *the moment you look at X, you cease to see its meaning.*"² When we concentrate our attention upon the universe itself, then, it gets shut up in itself, so far as our understanding of it is concerned, and thereby loses the range of depth in which its meaning as a whole is to be found. Even as a harmonious intelligible whole the universe can provide no explanation of its own inherent rationality. If we are to recover the meaning of the universe, and meaning of the universe as a whole, we must learn again to look beyond the universe, or look through the universe, to its transcendent ground in the uncreated Rationality of God.

This is certainly how the mediaeval world treated the created world, for they found the meaning of the whole of human life and thought in an other-worldly orientation. The arts and sciences were all deeply informed by a distinctive slant given to the classical attitude of mind through Augustinian metaphysics and eschatology, in accordance with which nature acquired significance only in so far as it provided the sensible medium in which eternal realities could be reflected and through which they might be discerned and enjoyed by men and women in their earthly pilgrimage to the heavenly city. This was a "sacramental universe" in which things outward, visible and temporal were regarded as signs of things inward, invisible and eternal. But this way of treating the natural order, only as a symbolic medium enabling the human mind to take off into a supernatural realm beyond, together with the radical dualism between the sensible and the intelligible orders of reality that lay behind it, had the effect of damaging the status and demoting the meaning of the creaturely world which it had been given in the Christian reconstruction of Greek thought. If we are to be true to those original Christian foundations, the meaning of the universe as a whole in its semantic orientation toward the eternal cannot be bought at the expense of the meaning which the universe has been given in itself precisely at its creation at the hands of God. Some way of bringing these two orientations together must be found, that is, in a profounder synthesis than any which informed the mediaeval world.

The difficulty we have to face, however, is that the all-important distinction between the creation and the Creator, which since the Reformation has been emancipating contingent being for empirico-scientific investigation out of itself, untrammelled by final causes or *a priori* dogmatisms, has been infiltrated by a secular form of Augustinianism to such an extent that there has resulted not only a deistic disjunction between God and the world but a further split within intra-mundane meaning between signs and things signified, so that false ontologies have arisen to bifurcate the formalisms we develop in thought and speech from the inherent structures of the created universe. This has everywhere become manifest in recent years in the bizarre contraposition between objectivism and subjectivism of various kinds, each side of the split apparently provoking the other to further extremes. It is in this light that one must regard the ambiguous plight of current theology: on the one hand, its defeatest retreat from externality into existentialism, and on the other hand, the "new externality" dialectically demanded by the "new inwardness" such as one finds in what is called "political theology". Caught in the chaos of man's alienation from the universe and in the schizoid state of affairs that opens up in his mental existence as ideas and events, and ways of understanding and explaining, fall apart, a "new theology" is found trying desperately to create meaning for mankind, now in this way and now in that, but all in vain, for the substitute ontologies intended to sustain meaning inevitably keep on collapsing on themselves.

If all this is to be avoided, we will have to get well behind the fatal dichotomies of which we have been speaking and ground our thought upon the creative relation between God and the universe in which he not only gives it its contingent being and confers upon it its own natural intelligibility, but so upholds and sustains it from below in its relation toward himself that he provides the creation with that unceasing reference beyond itself in which all its meaning is ultimately constituted. There is here no hiatus between the intrinsic and the extrinsic meaning of the universe, for it is precisely in virtue of its contingent condition, what it is in itself, that it points beyond itself. Since we ourselves are members of the universe, it is only within that contingent and semantic reference of the universe to the Creator that we may develop knowledge of God that is, within the space and time which God has brought into being with the universe as bearers of its rational order and through which he makes himself known to us and summons us to intelligent relation toward himself. Any attempt to explicate knowledge of God outside of or apart from those structures of space and time is inevitably and essentially irrational. We cannot know God apart from the way in which he interacts with the world he has made or apart from the way in which we are constituted his creatures within that world. If the creation of the universe out of nothing means that we may have knowledge of universe only out of its contingent processes, it also means that it is only from within the same universe and through the medium of its contingent rationalities that we may articulate the knowledge God gives us of himself, even though he infinitely transcends the universe. Since it is only within the ontological and referential relations of the universe to God that we may think and speak of God, there must be a close coordination between theological concepts and physical concepts: which is, after all, the inescapable implication of the Christian doctrines of creation and incarnation and the inseparable relation between logos and being which they establish. This being the case, an essential place must be found for so-called "natural theology", if only out of recognition of the fact that the interaction of God with the world grounds our conceiving of him within the relation of God to the world and of the world to God.

In view of this it is rather curious that natural theology seems to have flourished only in times when a cosmological dualism dominated thought and to have partaken of that dualism. In fact it is that kind of natural theology that is traditionally known as "natural theology". Two periods have been outstanding, that beginning with the twelfth century when the great mediaeval natural theology took its rise, and that beginning with the seventeenth century out of which there came, especially after Newton and Locke in the next century, the natural theology of English deism. There were of course real differences between them. Mediaeval natural theology was elaborated within a dualism in which the creaturely world was largely respected as a symbolic realm directing thought away to supernatural realities: nature was impregnated with meaning through semantic indications or intimations of a transcendent rationality beyond. That is why the abstractive processes of knowledge were held in check. The natural theology of English deism, however, was implicated in a dualism forcefully projected from the developing structures of natural science. It certainly had its roots in the new way of regarding the natural world in order to understand it out of its own contingent processes, *acsi deus non daretur*, but it was conditioned by the new dualism between absolute time and space and the events that take place within their all-embracing envelope, which in its turn gave way to a disjunction between God and a mechanistic universe so characteristic of German deism. This way of thinking was grafted on to the revival of Augustinian dualism evident in the Cartesian-Kantian development of European thought; but now abstractive processes could not be held in check so that when the critical reason questioned the validity of the bridge which natural theology had tried to throw between the sensible and intelligible realms, deism tended to give way to agnosticism, and even to atheism when the balance of thought swung over to a technological mode of rationality.

In both these periods, however, natural theology was pursued as an independent conceptual system, claiming to have its value precisely in that independent status, as a sort of *praeambula fidei*, antecedent to positive theology, fulfilling a mediating and apologetic function. "Natural theology" of this kind attempts to reach and teach knowledge of God, apart altogether from any interaction between God and the world, and proceeds by way of abstraction from sense experience and inferential and deductive trains of reasoning from observed or empirical facts. Again there are differences in the ways in which mediaeval and modern thinkers have carried out this programme. But they have all tried to establish some sort of bridge whereby thought can be invited to move inductively from observational experience to God. Yet in spite of the most prodigious efforts in mediaeval and modern times, a natural theology of this kind has not been able to gain the universal recognition which it desired or needed. It seems to have failed to understand the referential relations of forms of thought and speech to being, or of concepts to experience, for in one way or another they tended to be transposed into logico-deductive or logico-syntactical processes, with the result that when these constructions collapsed there opened up in late mediaeval as in modern times an unbridgeable chasm between thought and being or between ideas and events. That is to say, natural theology of this kind represents a desperate attempt to find a *logical bridge* between concepts and experience in order to cross the fatal separation between the world and God which it had posited in its initial assumptions, but it had to collapse along with the notion that science proceeds by way of abstraction from observational data. At its heart, as it would now appear to be the case, there lay a form of "logical empiricism" which was not able to stand up to the rigour of logical or empiricist analysis, but which shared with empiricism the positivist notion of knowledge that has now fallen before the advance of pure science. No doubt the weakness first became apparent in the way in which natural theology tried to relate the thought of God to sense experience, but when basically the same weakness appeared in positivist science in the way in which it sought to derive concepts from observations, that whole way of thinking met its end.

All this must not be taken to mean the end of natural theology, however, but rather its need for a radical reconstruction through a profounder way of coordinating our thought with being. What is involved here may be indicated by drawing an analogy taken from modern physics and the problem it has had to face in coordinating geometrical concepts and experience and which it came to solve by taking its cue from a Riemannian geometry of space-time. Euclidean geometry is pursued and developed *a priori*, as an independent science on its own, antecedent to physics, but is then found to be finally irrelevant to the actual structure of the universe of space and time. Everything changes, however, when geometry is introduced into the material content of physics as a four-dimensional physical geometry, for then it becomes what Einstein called "a natural science" in indissoluble unity with physics.[3] So it is with natural theology: brought within the embrace of positive theology and developed as a complex of rational structures arising in our actual knowledge of God it becomes "natural" in a new way, natural to its proper object, God in self-revealing interaction with us in space and time. Natural theology then constitutes the epistemological "geometry", as it were, within the fabric of "revealed theology" as it is apprehended and articulated within the objectivities and intelligibilities of the space-time medium through which God has made himself known to us. As such, however, natural theology has no independent status but is the pliant conceptual instrument which Christian theology uses in unfolding and expressing the content of real knowledge of God through modes of human thought and speech that are made rigorously appropriate to his self-revelation to mankind.

It is the coordination of the empirical and theoretical

components in that knowledge that is important, in a mutual relation in which they are neither confused with each other nor separated from each other, but in which the theoretical components serve the disclosure and understanding of the empirical. The problem to be handled in a movement of this kind has been well stated by Einstein. "The reciprocal relationship of epistemology and science is of a noteworthy kind. They are dependent upon each other. Epistemology without contact with science becomes an empty scheme. Science without epistemology is — insofar as it is thinkable at all — primitive and muddled. However, no sooner has the epistemologist, who is seeking a clear system, fought his way through to such a system, than he is inclined to interpret the thought-content of science in the sense of his system and to reject whatever does not fit into his system. The scientist, however, cannot afford to carry his striving for epistemological systematic that far. He accepts gratefully the epistemological conceptual analysis; but the external conditions, which are set for him by the facts of experience, do not permit him to let himself be too restricted in the construction of his conceptual world by the adherence to an epistemological system."[4] This certainly holds good for the way in which the theologian must keep in balance the epistemological structure which he cannot do without and the material content of his knowledge which he seeks to bring to conceptual order and articulation.

If we reject a deistic disjunction between God and the world, which we are bound to do, natural theology cannot be pursued in its traditional abstractive form, as a prior conceptual system on its own, but must be brought within the body of positive theology and be pursued in indissoluble unity with it. No longer extrinsic but intrinsic to actual knowledge of God, it will function as the necessary *intra-structure* of theological science, in which we are concerned to unfold and express the rational forms of our understanding that arise under the compulsion of the intelligible reality of God's self-revelation. As such, however, it is open to conceptual philosophical analysis, although in the nature of the case it cannot be isolated as an independent logical structure claiming to constitute an indispensable precondition for theological inquiry.[5] If in the relation of geometry to physics, as Einstein pointed out, it was forgetfulness that the axiomatic construction of Euclidean geometry has an empirical foundation that was responsible for the fatal error that Euclidean geometry is a necessity of thought which is prior to all experience,[6] theological science ought to be warned against the possibility of regarding natural theology in the heart of dogmatic theology as a formal system which can be shown to have validity on its own, for that would only serve to transpose it back into an *a priori* system that was merely an empty scheme of thought. In this transformed way we may speak of natural theology as constituting a necessary condition but not, as Henri Bouillard has rightly pointed out, a sufficient condition for theological knowledge.[7] If therefore it is to serve its function within theology, natural theology cannot be treated as being complete and consistent in itself, but only as attaining consistency within the empirical conditions of actual knowledge of God, and therefore as an essentially *open, pliant structure*, involving basic concepts that are decidable only on other grounds or at another level of thought.

In the light of the foregoing discussion it must be said that traditional natural theology, however laudable the intentions that may have been behind it, had the effect of providing a prior framework of thought within which the material content of actual knowledge of God mediated through divine revelation was interpreted and moulded without respect to the empirical conditions under which it arose, and was thereby seriously distorted. In attempting to establish a logical bridge between concepts and experience, or idea and being, and thereby to reach out inferentially to God, it was committed to a logical formalisation of the empirical and theoretic components of knowledge of God which reduced them if not to a single conceptual system at least to the same logical level. This position was considerably reinforced in mediaeval thought by the unquestioned assumption that to think scientifically was to think *more geometrico*, that is, on the model of Euclidean geometry, and it was reinforced in later thought as it allowed itself to be restricted within the logico-causal connections of a mechanistic universe.

Since these ways have met their end, how do we proceed? Clearly some sort of bi-polar relation between natural theology and revealed theology is demanded, in which we do not operate with an identity in which natural theology would be reduced to revealed theology, but with an inner correlation between them similar to that in which geometry and physics are dependent on each other, but in which geometry, while incomplete and ultimately inconsistent without integration with physics, still remains a distinct discipline or set of disciplines which can be legitimately pursued in a form which is methodologically or notionally bracketed off from the material content of physics for purposes of logical clarification. A geometry completely detached from physics through idealising abstraction from the actual spatial and temporal structures of the universe could be made internally consistent if it were reduced to an empty tautological system, but only at the expense of being rendered irrelevant and useless for physical knowledge of reality. Likewise a natural theology completely severed from revealed theology through idealising abstraction from the spatial and temporal correlates of God's interaction with us in the universe, could be made internally consistent if it were reduced to an empty tautological system, but only at the expense of being rendered irrelevant and useless for real knowledge of God. It must be frankly admitted, therefore, that if natural theology is to be the object of independent consideration and analysis as a discipline on its own, that can take place only through being *artificially* bracketed off from the material content of actual knowledge of God, and could be accepted only as a *temporary* methodological device for purposes of clarification. Even then, however, and this also must be frankly admitted, the natural theology which we methodologically bracket off in this way, still retains the imprint of its empirical origins and foundations, which means that in our clarificatory analysis of it we cannot in truthfulness forget its correlation with revealed theology. It is, then, only with a profound recognition of its epistemological polarity with revealed theology that we may turn to consider the status of natural theology itself, without letting it lapse into the position of an antecedent and independent conceptual system on its own. That is what we seek to do with a proper natural theology in this chapter, leaving to the next chapter the other pole of the correlation in the rigorous science of revealed theology, although we will not be concerned in it, or in any of the later chapters, with its material content so much as with its structure as a positive or dogmatic science in its own right.

It may help us at this juncture to borrow a rather telling analogy from Michael Polanyi,[8] taken from what happens when someone puts on spectacles that invert his vision, making him see things upside down or, as the case may be, on the right instead of on the left and vice versa, thereby disrupting the coordination of his perceptual and mental images. For some time he is badly confused and helplessly disorientated, and it takes about eight difficult days before he can readapt himself to the world about him sufficiently to interact with it in a rational, objective way. The ability to see things the right way round is not gained by applying explicit rules but only through a tacit process of learning — which is Michael Polanyi's interest in the experiment. He points out, however, that readjustment is not achieved through a change in the perceptual image but through a change in the mental image. Proper vision and objective behaviour are restored only through a *conceptual reform* in which perceptual and mental images are realigned with one another, that is, one in which empirical and theoretical factors are fused together in an epistemically unitary outlook upon the world in accordance with the actual way in which they inhere in one another in the real world independent of our observations.

That is the kind of conceptual reform that is now taking place on a comprehensive scale, with the fundamental shift in our understanding of the universe that has come with the advent of relativity theory and its unification of ontology and intelligibility in scientific knowledge. We are evidently undergoing a slow reorganisation of our basic conceptions which restores us to ways of inquiry and understanding in which aspects of reality which had been torn apart through dualist and abstractive modes of thought are allowed to remain together in their natural unity and to disclose their meaning to us under the force of their intrinsic intelligibilities.

We now need a similar conceptual reform in our outlook upon the universe as a whole, after the myopia that overtook our vision during centuries of deism and secularism, and the loss of meaning that resulted from the foreshortening of semantic perspective which they brought about, that is, a habit of looking at the universe in such a way as to cut off its signitive or referential relations beyond itself. What we need is such a shift in the focus of our vision that, instead of looking at the universe in the flat, as it were, we look at it in a multidimensional way in which the universe as a whole, and everything within it, are found to have meaning through an immanent intelligibility that ranges far beyond the universe to an ultimate ground in the transcendent and uncreated Rationality of God.

From the perspective of divine revelation, of course, the required shift in vision and conceptual form will derive from the Word or Logos of God as through the incarnation it intersects contingent being and intelligibility and gives them a unifying semantic reference which terminates on God himself. From the other pole of this relationship, however, where our attention is directed in natural theology, it is the actual contingent nature of being and of its inherent intelligibility that is all-significant, since precisely as *contingent* creaturely *being* and *intelligibility* require a sufficient ground and reason beyond themselves in order to be what they actually are.[9] In other words, contingent being and contingent intelligibility in virtue of what they actually are constitute a rational question requiring a rational answer. In order to open out discussion of this state of affairs, let us return to a point raised in the preceding chapter regarding the problems that arise when *ousia* and *logos*, or being and intelligibility, are torn apart.

As we have already noted, it was the dualist outlook, both cosmological and epistemological, resulting from that bifurcation that occasioned the distorting short-sightedness in our vision, and it was within that dualism that traditional natural theology arose in its damaging, independent and prescriptive form. The situation is very different, however, if we start from the unity of being and intelligibility, as we are now bound to do in view of the changes in the foundations of knowledge forced upon us through relativistic physics. In this event we adopt an approach to the nature of things in which we let reality disclose itself to us in its primordial state, so that we may apprehend it in accordance with its inherent structures and read its meaning as we penetrate into its own power signification. If *ousia* and *logos* are allowed to operate in this kind of cohesion, if the being of the universe in its inherent intelligibility, in virtue of what it actually is, is allowed to constitute a rational question requiring a rational answer, then clearly the fields of inquiry in ontology and natural theology overlap. This is not to say that they can be reduced to one another, but that there is a very close relation between them, not least in respect of the way in which each proceeds. It may be instructive, therefore, in the first instance to consider simply our approach to being as such — *ens inquantum ens*.

It is imperative to leave behind any idea that "being" is what we "know" by objectifying modes of apprehension or thought which we found to be so characteristic of "the modern attitude of mind". Being represents rather what we know in every field of experience as we let what we know be what it actually is without changing or modifying it. This applies to persons as well as things, to subject-beings as well as object-beings, for both may be objects of knowledge in the proper sense, and in the proper sense constitute objective realities or beings in their own right.

An approach to being along this line has been well set out

and cogently argued by Julian Hartt in a little book in which he has drawn attention to three basic features of objective being: its power to appear, to act from interior principle, and to be signified.[10] To be an object of knowledge something must have a natural power of being to appear to another being, to show itself or communicate something true of itself. "Communication, so understood, is the pattern or patterns a thing has in relation to other things in which its true nature is exhibited."[11] Objective being is also characterised by "the power to act from interior principle". "When we are dealing with an objective thing we know we are dealing with something which has its own source of activity within itself, at least relative to the knower. We know also that this source is concealed from us: its power to act is 'interior' to itself. We know what it is specifically only so far as it 'appears'."[12] Objective being has the further feature of being able to be signified, to be what is intended or meant in our knowledge of it or discourse about it. Signification, intention, and meaning, however, are to be taken in their objective sense, referring not to some purpose or intent of ours but to an objective state of affairs in being which can be attended to as having a certain character or nature. All being admits of signification in this way, but precisely as objective being there is far more to it than we can know. "Our finite minds cannot match the power of signification in being itself."[13] Along with a showing forth there remains a concealment.[14]

This account of being as the object of knowledge is complemented by another account which Professor Hartt gives of being, as the subject of knowing.[15] Subjective being is considered as the power to be affected, to grasp an interior principle of activity, as the power of intent and the power to judge, and through the correlation of these he shows that being itself is the ultimate subject of attribution and as such is the referent or ultimate ground with which we are concerned in questions of truth and falsity.[16] Propositions and structures of meaning are true only when they manifest the intention of being itself, but "being itself is power or powers beyond all structures of meaning within our grasp".[17] We will not follow Julian Hartt's train of thought further, for it should now be evident that if we let our minds approach being in this way, in accordance with its own interior principles of activity and its own powers of signification and communication, instead of developing objectifying and cramping modes of thought, they develop objective modes of thought correlated with the ultimate openness of being and its semantic reference beyond, which will not be restricted to the stipulations which we so often carry in our specific questions.

It is a similar approach that is found in the thought of Martin Heidegger, who is concerned to break up the rigid structures of language and existence in order to let being itself show through in its freedom and primordial reality. Heidegger is rather more specific than Hartt as to the relation between *logos* and *physis*. Logos is the natural force of being to manifest itself, to come out into the open and show itself in its own light. Logos is not itself the locus of truth, but is the manifesting of the reality of things or their "unconcealment" (*alētheia*).[18] It is significant that here we do not need and do not operate with intermediary representations or sense-data, the sort of thing that seems forced upon us whenever we allow logos to secede from being, for what we are concerned with here is the showing of reality itself through to us. This is why Heidegger has devoted so much effort to analysing "existence" (*Dasein*) in order to destroy the false ontologies that keep on cropping up through the hardening of substitute-symbolisms, so that the whole focus of attention may be directed upon being in the full and proper sense (*Sein*). Heidegger's own movement of thought, however, tends to be stultified and distorted, for he lowers the horizon of inquiry (*Fragehorizont*) in a strangely arbitrary way, so that in the last resort he is forced to break out by taking a leap into nothing, instead of letting his mind fall under the power of the inherent signification of being and its reference beyond itself. But far from opening up the reality of being, as Heidegger hoped, the leap of thought into nothing can only fall back into the relativities of existentialism.

When reality is allowed to unveil itself, however, in its own inner intelligibility, our thought is thrust up against the truth of being in such a way that it is sustained by an objective signification and does not fall back into the dark whirlpool of man's self-understanding. Not only do we grasp the truth of intelligible being out of the depth of its own reality, but we let it interpret itself to us as we develop appropriate structures of thought under its pressure upon us, i.e., as we respond to that which is *not* of our own making and which acts with categorical force upon our minds. Here too we find that explaining and understanding do not fall apart, as they do when logos and being become disjoined, for they operate together as we penetrate into the inner relations and significations of things and allow them to set up their own laws and meanings in our grasping of them. To express this the other way round, what we apprehend like this in the truth of its own being *proves itself* to us by bringing our minds under an imperative obligation which we cannot rationally resist. This is what happened, for example, when St. Anselm, breaking free from the psychologism of St. Augustine, could speak of truth not only as that which is what it is and as it manifests itself, but as what it *had to be* in his understanding or conceiving of it, for he found himself together with the truth of created being under compulsion from beyond. It was in this way that the so-called "ontological argument" *forced* itself upon him from the side of what he had to acknowledge as the Supreme Being or Supreme Truth.[19] That is not to be understood as the force of some logical argument or deductive or inductive train of ideas, but as the force of Reality itself (*necessitas*, as St. Anselm called it, its impossibility of being otherwise) in which truth and being are indissolubly one.

Before we go on to draw out the implications of this approach for natural theology, we must pursue a little further its bearing upon questions of truth and falsity. Within this orientation the true is that which allows being to stand out before us in its self-identity and persistence, to impress itself in its reality, unity and structure upon us in

STATUS OF NATURAL THEOLOGY 49

such a way that our minds come under the compulsion of what it is in itself and its interior relations or signification. Whenever logos relates to reality in this way the forms of thought or the patterns of meaning which we develop may be true or false. If they serve the uncovering of being, let it show itself or come to view or stand out in its reality, they are true, but if they obscure being or distort its showing forth by imposing themselves upon it as an alien structure of meaning, in which the human logos thereby makes itself the measure of things, they are false. Whenever we operate with the unity of logos and being, however, reality itself must be the ultimate judge of what is true or false, for the truth or falsity of our concepts or statements lies not in them but in their bearing upon reality. It is only where the relation of logos and being with one another is damaged or refracted that we find truth being conceived in terms of congruence or adequation.

That is to say, whenever we try to develop our knowledge by abstractive processes some type of correspondence view of truth is forced upon us by the prescinding of form from being, and then as the form assumes an independent status of its own, it is some standard of inner consistency or a coherence view of truth that becomes uppermost. In the process of this development the damarcation/falsifiability test advocated and discussed by Karl Popper has its point. But the question must be raised whether a logico-deductive testing of scientific theories assumes that the relation between our concepts and being can be specified in a clear and determinate manner, and further whether there is not here a "hang-over" from the positivist assumption of a logical bridge between concepts and experience? Certainly testing for a consistent internal structure is always in order, while applicability to existence and therefore demarcation tests are a *sine qua non* of verification in scientific theory, but does not the falsifiability test imply a correspondence view of truth and therefore apply only where logos is somehow disjoined from being? If the falsifiability test is to escape the charge of working with a logical bridge between concepts and experience in the testing of our theories, all that it can claim to show is that a theory is not necessarily true, not that it is false. If, however, we reject the notion of a logical bridge between our concepts and experience in the verification of our concepts, just as we reject such a bridge in the rise of these concepts in discovery, the decisive place in verification, as in discovery, must be given to that coordination between concepts and experience which is not susceptible of formalisation or logical determination. Hence, as Polanyi rightly asserts, it is always reality itself which must be our final judge in submission to which the personal judgment of the scientist is offered.[20]

It is not with logical necessity that we are primarily concerned in testing the truth or falsity of scientific concepts and theories, but with ontic necessity, the impossibility of what is the case being otherwise, and therefore the necessity in our conceiving of it in accordance with what it is and not otherwise. That is why in verification we seek for some sufficient reason grounded in the *esse* of what is known or being investigated, through which it may become evident whether there is a *nec-esse* relation between our knowing and what we know — i.e. one that cannot be otherwise. And so what we try to do in establishing our theories is to ground thought upon immanent relations in being in virtue of which things are what they really are, that is by rightly referring to and serving the conditions of reality beyond themselves, not by trying to connect thought logico-deductively with being. Verificatory procedures have to do with states of affairs or conditions of reality, so that true theories must always terminate upon being, not upon relations of ideas to which relations of matters of fact have been reduced or even upon relations of thought which are made to suppost for relations in being.

It is important to note, however, that while this applies to every field of being, it does not apply to *being itself* in the same way as to particular beings or fields of being, where reference to some specific set of conditions or some particular states of affairs enters into the verification or falsification of claims to knowledge, but only where there is reference to the ground of being which encompasses all states of affairs. It is the case, of course, that in every particular instance or field of being, where we direct our questions to things in order to allow them to disclose themselves and at the same time check our apprehension of disclosures in terms of the sets of conditions of reality that come to view, our thought terminates upon being itself as its *final* objective, but so far as knowledge of being itself is concerned verification of our apprehension of its disclosure must presuppose being itself from the very start as well as terminate upon being itself at the end. Hence while we are concerned here too with processes of knowing and verifying in which we allow reality to come to view and be the final judge of our knowledge, we are thrown back in the most absolute sense upon the self-evidencing of reality as the ground upon which we make statements about it, and with reference to which even false statements are made, for after all there must be something there if we are to speak of it as false. What inquiry and verification do is to cleanse our conceptions of those elements deriving *from ourselves* which obstruct and distort the manifestation of being itself. Here, then, we can get no help from knowledge of particular or determinate realities, for the specifying modes of thought to which they give rise are resisted and relativised by being itself. We can only try to resolve any conflict between conception and disclosure face to face, as it were, with being itself, and so try to match our apprehension of it with its own power of signification, while remembering the point made by Julian Hartt that "being itself is power or powers beyond all the structures of meaning within our grasp".[21]

We have been discussing the *modus cognoscendi* which we develop in ontological inquiries where we hold to the primordial unity of *physis* or *ousia* and *logos*, because it had become clear that the fields of inquiry in ontology and natural theology overlap. But now we must ask what this implies for a natural theology in which we operate not by abstraction from sense experience but through consideration of being in its inner intelligibility, and in which we have to do with the natural pole of the correlation between theological concepts and physical concepts. What happens when we let our minds fall under the intrinsic signification of contingent being and its reference or openness beyond itself?[22] In all natural science we are concerned to lay bare the intelligible structure of the universe, and try everywhere to grasp reality in its own rational depth, but everywhere we are faced with questions as to the very existence of such a universe, and as to the immanent rationality embedded in that which exists. Why is there a universe and not nothing? What is the reason for this state of affairs, the existence of a universe that is accessible to rational inquiry? Yet the universe does not carry in itself any explanation for this state of affairs, and even the rationality embedded within it is not self-explanatory. This is certainly understandable, for contingent being cannot explain itself, otherwise it should not be contingent. Nevertheless it does have something to "say" to us, simply by being what it is, contingent *and* intelligible in its contingency, for that makes its lack of self-explanation inescapably problematic, and it is precisely through that problematic character that it points beyond itself with a mute cry for sufficient reason. What the intelligible being of the universe has to "say" is thus something which by its very nature must break off in accordance with the utterly contingent existence of the universe. This may be expressed more positively: the fact that the universe is intrinsically rational means that it is capable of, or open to, rational explanation — from beyond itself.[23]

This is the direction in which our thought is carried by the very advance of actual knowledge in natural science, for the more the intelligible universe becomes disclosed to our inquiries, the more it is revealed to be one that stretches out indefinitely beyond our grasping and articulation of it, so that we are forced to think of it as having a vast range of intelligibility which we can apprehend only at relatively low levels. We are baffled not only by the nature but by the existence of an intelligibility which we cannot fully penetrate, but at the same time we stand in astonishment at the fact that the concepts that arise in our minds as we intuit and grasp the reality of things are rationally correlated to it, even though they fall far short of it, and although we cannot explain how this is so. A double fact evokes our wonder: the openness of the structure of the universe to our rational investigation, and the openness of our knowing to the intelligible nature of the universe. This is what Einstein used to speak of as the religious awe in which he was left by the vast comprehensibility of the universe. The fact that it is comprehensible at all to us is a miracle, indeed the most incomprehensible thing about it.[24] Now it is just this state of affairs that *suggests*, or directs us to, a transcendent ground of rationality as its explanation. It is the objective depth of comprehensibility in the universe that projects our thought beyond it in this way, for complete enclosure within its own immanence would contradict the nature of that comprehensibility and threaten the universe with ultimate meaninglessness. The unbounded range of its comprehensibility refuses explanation merely in terms of reasonable relations between particular states of affairs within the universe. To be inherently reasonable the universe requires a sufficient reason for being what it is as an intelligible whole.

Thus we find ourselves in a situation where the intelligibility manifest in and through the universe seems to lay hold of us with a power which we cannot rationally resist. It is part of our rationality that we act under the compulsion of the nature of things and assent to it in a positive way, but here there is a relation of transcendent reference which catches up on us and requires of us the same kind of assent, for somehow we are already committed to it in the correlation of our mental operations and the open structures of the universe of being. While it is certainly true that the semantic reference of the intelligible system of the universe breaks off and can only point brokenly beyond, so that the intentionality it involves in virtue of its contingent nature does not terminate upon an identifiable rational ground, nevertheless we are aware of coming under an imperious constraint from beyond which holds out to us the promise of future disclosure and summons us to further heuristic inquiry which it would be irresponsible of us to evade. It is important to note, however, that the force of this constraint is inseparably bound up with the obligatoriness of being and its immanent rationality that bear upon us in the universe, and therefore with the cataleptic consent which we are bound to yield to the given reality of things beyond our conceptual control or manipulation.

This is a way of thinking, it must be admitted, that cannot readily be appreciated from the side of positivist science, for, "positivist" though it is, its detachment from being precludes the kind of positive assent demanded of us face to face with being. Through its modes of procedure, which Popper speaks of as "observationalist" and "instrumentalist",[25] it limits itself to operational and pragmatic, and in that sense, merely provisional statements which have no direct ontological reference. Positivism defines scientific theory in terms of abstraction from observational data and the formation of convenient functional relations between observational concepts, that is, as "a working hypothesis" and no more. All metaphysical claims, that is, all claims bearing upon being, are here excluded, for no thought is entertained of grasping reality in its depth. This positivist notion of science stands in rather sharp contrast to Einstein's conception of science, in which we are concerned in the development of scientific theories to penetrate into the comprehensibility of reality and grasp it in its mathematical harmonies or symmetries or its invariant structures, which hold good independently of our perceiving: we apprehend the real world as it forces itself upon us through the theories it calls forth from us. Theories take shape in our minds under the pressure of the real world upon us and are the cognitive instruments through which we speculatively penetrate into the objective order of the intelligible universe and let it become revealed to us. Theories are certainly to be modified, or refined, like lenses, in the light of what is disclosed, but they are essentially disclosure models which claim recognition for what they serve to reveal.[26] However, as that objective order becomes revealed to us, even in some measure, and we let our minds fall under its compulsive force, we cannot think that it might be otherwise. What we apprehend and articulate in this way, e.g. through what we call a natural law, is an identifiable datum which we must affirm as such in positive assertions about it. This is the inescapable "dogmatic realism" of a science pursued and elaborated under the compelling claims and constraints of reality.

In discussing the nature of the assent which scientific concepts and statements involve, Michael Polyanyi has developed an illuminating comparison between a noise and a tune.[27] According to information theory a noise is a random series of sounds which conveys no meaning or message, but a tune is a distinctive set of signals which are identifiable in an ordered sequence and which can be read, therefore, as a sort of message. Ideally any single message is represented by only one configuration of signals. But the opposite holds good for noise, for no particular significance attaches to a set of random sounds, and it makes no difference what sounds it comprises: they could just as well have been different. This is precisely what does not apply to an identifiable sequence of sounds which is recognised as having a meaning or conveying a message, for we cannot think that it could have been otherwise. In apprehending it we find ourselves under the compulsion of its meaning and, as Polyanyi says, we have to affirm it with universal intent. That is to say, we do not have to do simply with a functional arrangement for certain purposes of our own, but with an objective structure, invariant for any and every observer.

When we come up against the universe in that profound dimension of its inherent intelligibility we find ourselves, as we have already noted, in a situation in which we are committed and obliged to assent to it. But further, we are taken in command by an intelligible reality that transcends our explicit formalisations of it and indeed any ultimate explanation from the side of our knowing of it. It is this unbounded range of intelligibility that is so awesome and so wonderful. Now this means that we cannot rationally break off our relation to the intelligibility of the universe at some arbitrary point of our own choosing; otherwise it would not be existence or reality that we are determined to apprehend.[28] This is why scientific inquiry cannot come to a halt at any point we want, but must go on questioning its questions in order to let reality disclose itself to science indefinitely. The capacity of man for this kind of indefinite, unlimited, unbounded inquiry represents that which from his side is correlated to the intelligibility that reaches out indefinitely beyond him and which cries out for, and manifests itself as capable of, explanation in relation to some transcendent source and ground of rationality.

Even on this merely "natural" level of human inquiry, then, the fact that the intelligibility of the universe is not self-explanatory and the fact that its astonishing unity and self-identity persist through all its changing processes, would seem to *suggest* that there is an *active agency* other than the inherent intelligibility and harmony of the universe, unifying and structuring it, and providing it with its ground of being.[29] In view of this it is not surprising that a natural theology, which is aware that it cannot operate with any logical bridge between concepts and being, should try to develop some postulatory or axiomatic mode of thought in order to find a way through the contingent intelligibilities of the universe to the point where it might be discerned whether such an active agency would throw any light upon the problematic intentionality of the universe by supplying it with its transcendent referent, and thus sustaining the universe in its ultimate meaningfulness by grounding it beyond itself. There are people, however, perhaps many, who are unable to follow this line of thought, but the question must be asked whether that is due to a difficulty inherent in a movement of relating the universe to an active agent or to the fact that they have been conditioned against it by working so long with essentially impersonal and observational models of thought deriving from the Cartesian bifurcation of subject and object and developed through severely abstractive processes of theorising. If man is considered only as "thinking thing" poised upon himself over against the world out there (*res cogitans* over against *res extensa*), then the world can be brought within the knowledge of the detached subject only by way of observing phenomena, accounting for them through determining phenomenal connections, and reducing them to rational representation. Thus the "world" is that which is constructed out of the states of man's consciousness, not something with which he interacts as a personal agent: it is merely the subject of his objectivist and objectifying operations. It was of course this way of thinking that gave rise to the mechanistic view of the universe within which man has found his personal existence threatened through the tyranny of rigid, impersonal concepts, but that result was implied in the assumptions of its starting point.

It is evident that a Cartesian, observationalist and abstractive way of thought inevitably detaches any thought of God from thought of the world, as well as any thought of man as active agent from thought of the world, so that on this model of thought we cannot arrive at, or we inevitably exclude from our thought, any idea of God as interacting with the world and interacting with man in his personal and historical existence within it. But it is *action*, in which we personally behave in accordance with the nature of things around us, that connects man and the world in a way that overcomes the detached relation between man and nature. Hence a recovery of the concept of the human being as personal agent, actively related to the world of things and persons around him, erases the radical dualism upon which the old model of thought depended (i.e. the model built up from the concept of man as a detached observer over against inert, determinate being), and produces a new orientation of mind to the universe in which the idea of a God who interacts with us and our world is not automatically excluded.[30]

More positively, if we think of man as actively at work in the universe, in the interrelations of people with one another and the world around them, and from the dynamic correlation of intelligibility to intelligence, to which it gives rise, develop a model of thought in terms of *active agency*, in place of the old observationalist and impersonalist models, then it will surely *suggest* even more strongly that the universe is what it is under the agency of a transcendent Being. Certainly if we look at contingent being, abstracted from our personal relations to it, as such, we can say no more than "suggests", for being as such by virtue of its contingent character cannot give any answer to our inquiry as to the ground beyond itself, but it does more than raise a question for it seems to *cry silently* for a transcendent agency in its explanation and understanding. It must be emphasised again, however, that no idea of a logical bridge can be contemplated here, no chain of inferential reasoning from the contingent to a "necessary" Creator is in place. But this is not to say that a deeply rational movement of thought is thereby excluded — if it were, all pure scientific thought would be excluded in the exact sciences, for even there it is impossible to operate with a logical or inferential coordination of concepts and experience. Nevertheless, just as there the intelligibility of the universe shows through to us and is accessible to our conceptual representations and formalisations, while retaining its inexhaustible and ultimately ineffable character in face of our apprehension of it, so here in their own unique way the Reality and Intelligibility of God may break through to us in ways we can recognise and apprehend without infringement of their transcendent character. Moreover, here we find our human being opened up and disclosed to us as there strikes at us through the blank face of the universe a mysterious intelligibility which takes us under its command in such a way that we feel we have to do with an undeniable and irreducibly transcendent reality which becomes intensely meaningful as the inward enlightenment of our own beings is correlated to it. It is understandable, therefore, that John

STATUS OF NATURAL THEOLOGY

Calvin could argue that it is only as knowledge of God strikes home to us in such a way that we come to know ourselves in a way we could not otherwise, that we are convinced that we really know God.[31]

On the other hand, if our thought along these lines really has to do with an active Agent who is the creative Source of the intelligibility of the universe, then we know him not because we succeed in penetrating through the intelligible structures of the universe to net him, as it were within our postulatory or axiomatic movements of thought, but rather because he actually interacts with us and the universe, constitutes himself the active Object of our knowledge, and discloses himself in a positive way to us as the created universe by virtue of its sheer contingency is quite unable to do. That is to say, knowledge of the living creative God is to be conceived as taking place within an empirical relation in which he so acts upon us as to provide us with the basic clues or proleptic conceptions without which our inquiry into God could not even begin.[32] Yet it is not with *discovery* that we have to do here, as in our inquiries into mute and determinate realities when we seek to let them "disclose" themselves to our questioning, but with *revelation* in which our seeking and inquiring are anticipated, prompted and supported by creative activity on God's part.

What we have been seeking to do in this chapter, of course, is to examine the formal structure of our human understanding of God, apart from the divine side of the bi-polar relationship which knowledge of God involves, methodologically and artificially isolating it for analysis by itself. Admittedly this is rather like converting four-dimensional geometry back into three-dimensional Euclidean geometry, or physical geometry back into *a priori* geometry! Actually, however, we may discern the rational structure of our understanding of God properly only within the coordination of what we are given to know and our knowing of it. This means that methodological bracketing off of one pole from the other must not be taken too seriously and certainly must not be allowed to harden, for as soon as it is reduced to a conceptual system on its own it becomes severely distorted, rather like what happens when we project a map of the round world on to a flat surface. We must be ready with our theological "transformation equations" to cope with the distortions and transpose any "results" of our analysis back into their original form. Hence any patterns of thought which we reach through a methodological bracketing of one pole from the other, cannot be reckoned to have any more than a quasi-validity within the artificial limits imposed, for proper analysis must take place within the empirical and theoretical coordination in which knowledge of God actually arises.[33]

It follows that what we have said in this chapter cannot stand alone, but must be taken together with the succeeding chapters, if it is to have its right force. A proper natural theology, we have argued, may be pursued only in indissoluble connection with revealed or positive theology, but then it is found to coincide with the epistemological intra-structure of our knowledge of God. No more than that latent epistemological structure, can natural theology be abstracted from the material content of actual knowledge of God and erected into a sufficient as well as a necessary condition for knowledge of God. Natural theology considered in itself alone is thus incomplete and only consistent if it is coordinated with positive theology, for it makes use of concepts and theorems which lack meaning and cogency in themselves but which may become meaningful and cogent when they are sublimated and interpreted from the level of divine revelation. Yet it is perhaps this open character of natural theology which gives it a continuing place in Christian thought, for it enables it to play a significant role in the coordinated levels of scientific theology without forcing upon it any distorting reductionism.

Hence as we proceed to discusss the positive pole of our knowledge of God we must constantly bear in mind the intimate and distinctive relation between epistemological structure and scientific subject-matter. The theoretical and empirical components of knowledge are coordinated in such a way as to serve its material content, but while the theoretical components by themselves tell us nothing, the empirical components by themselves remain opaque. We cannot argue at any point from forms of thought to objective structures in being, even though they arise in our minds under the pressure of those structures, but neither can we make any progress in grasping and interpreting those structures apart from the forms of thought to which they give rise in us. On the other hand, we do have to assume the reality and rational accessibility of those structures in testing or establishing the forms of our thought about them. This state of affairs holds good for every science where we operate within the correlation of our concepts with experience and seek to develop appropriate modes of inquiry through which we may penetrate into the intelligible realities of the field in question and develop at the same time appropriate modes of verifying them. Surely no other way is to be contemplated in theological science, although scientific rigour demands that we take the distinctive nature of the divine Object into account at every step.

NOTES

1. It was on this basis that Hugo Grotius developed his argument in *De iure belli et pacis*, 1.xff, i.e. *etsi deus non daretur*. *Acsi* seems to make more sense than *etsi* in this celebrated statement.
2. M. Polanyi, *Knowing and Being*, London, 1969, p. 146. "What happens is that our attention which is directed *from* (or through) a thing to its meaning is distracted by looking *at* the thing . . . To attend *from* a thing to its meaning is to interiorize it, and to look instead *at* the thing is to exteriorize or alienate it . . ."
3. A. Einstein, "Geometry and Experience", in *Sidelights on Relativity*, London, 1922, p. 27ff; *Ideas and Opinions*, New York, 1954, pp. 232ff. Cf. *The World as I See It*, London, 1935, p. 183: "Physical geometry is no longer an isolated self-contained science like the geometry of Euclid."
4. A. Einstein, in his reply to criticisms in P. A. Schilpp, *op. cit.*, pp. 683f.
5. See "The Problem of Natural Theology in the Thought of Karl Barth", *Religious Studies*, vol. 6, 1970, pp. 128ff; and also Henri

Bouillard, *The Knowledge of God*, London, 1969, from whom I have borrowed the term *intra-structure* (wrongly rendered in the English edition as *infra-structure*).
6. This way of expressing Einstein's thought is taken from V. F. Lenzen, "Einstein's Theory of Knowledge", in P. A. Schilpp, *op. cit.*, p. 369.
7. Henri Bouillard, *op. cit.*, p. 29.
8. M. Polanyi, *Knowing and Being*, pp. 144, 198ff.
9. Cf. L. Wittgenstein, *Tractatus Logico-Philosophicus*, 6.41, London, 1961: "The meaning of the world must lie outside the world. In the world everything is as it is, and everything happens as it does happen: *in* it no value exists — and if it did exist, it would have no value. If there is any value that does have value, it must lie outside the whole sphere of what happens and is the case. For all that happens and is the case is contingent (*zufällig*). What makes it non-contingent (*nichtzufällig*) cannot lie *within* the world, since if it did it would itself be contingent (*zufällig*). It must lie outside the world." I have replaced "accidental" and "non-accidental" in the Eng. translation by "contingent" and "non-contingent", which is truer to Wittgenstein's thought.
10. Julian Hartt, *Being Known and Being Revealed*, Stockton, 1957.
11. *Ibid.*, p. 13.
12. *Ibid.*, p. 14.
13. *Ibid.*, p. 16.
14. *Ibid.*, p. 16.
15. *Ibid.*, pp. 19–23.
16. *Ibid.*, pp. 24f, 30ff, 48ff.
17. *Ibid.*, p. 39.
18. Martin Heidegger, *Being and Time*, London, 1962, pp. 55ff, etc. Cf. also *Existence and Being*, London, 1949, pp. 132ff; and *An Introduction to Metaphysics*, Newhaven & Oxford, 1959, pp. 21, 102ff, 188ff.
19. St. Anselm, *Proslogion*, prol., *Opera Anselmi*, edit. by F. S. Schmitt, Edinburgh, 1946, vol. I, p. 93. The same point is also made by John Calvin, *Institutio*, I.16.1, where he makes use of Anselmian language: *quae sponte sese proferunt, et nolentibus ingerunt*. Cf. 1.4.1f.
20. M. Polanyi, *Knowing and Being*, pp. 119ff, 133ff, 172ff.
21. Julian Hartt, *op. cit.*, p. 39. This disparity between concepts and being need not alarm us, for it actually serves their denotative purpose in pointing away from themselves.
22. See E. L. Mascall, *The Openness of Being. Natural Theology Today*, London, 1971, The Gifford Lectures in the University of Edinburgh 1970–1971, in which a sustained and powerful argument is offered along the same line.
23. Cf. here *Divine and Contingent Order*, Oxford, 1981, which is devoted largely to this point.

24. A. Einstein, "Physics and Reality", *Out of My Later Years*, New York, 1950, p. 60f. Cf. P. A. Schilpp, *op. cit.*, pp. 284f, 365f. Also *The World as I See It*, p. 28: "The scientist... His religious feeling takes the form of a rapturous amazement at the harmony of natural law, which reveals an intelligence of such superiority that, compared with it, all the systematic thinking and acting of human beings is an utterly insignificant reflection."
25. Karl Popper, *Conjectures and Refutations. The Growth of Scientific Knowledge*, London, 1969 edit., pp. 21ff, 62f, 97ff, 107ff, 123, 127ff, 137f, 166ff, 173, 223, 226, 235, 245, 382, 408ff. Also *Objective Knowledge*, Oxford, 1972, pp. 64f, 69, 80, 194f, 262f.
26. See *Theological Science*, London 1969, pp. 241ff, 256ff, 286ff. For the changes in the meaning of "theory" see Hannah Arendt, *Between the Past and the Future*, London, 1961, p. 39f.
27. Michael Polanyi, *Knowing and Being*, p. 109.
28. Cf. A. Einstein, *The World as I See It*, p. 140: "It is existence and reality that one wishes to comprehend."
29. Cf. *God and Rationality*, London, 1971, p. 141. The fact that our understanding of the inherent intelligibility of the universe requires a close coordination between number-rationality and word-rationality supports the suggestion that here we have to reckon with a transcendental openness such as is amenable to word language, for which a personal rather than an impersonal model of thought would be appropriate.
30. The place of action and the model of active agency in science are recurring themes in the philosophy of John Macmurray. See *Reason and Emotion*, London, 1935; *The Boundaries of Science*, London, 1939; *The Self as Agent*, London, 1957, etc.
31. John Calvin, *Institutio*, I.1.1f.
32. Cf. the essential place of what St. John of the Cross called the *divine touch* in our intuitive and auditive apprehension of God. *The Collected Works of St. John of the Cross*, transl. by K. Kavanaugh and O. Rodriguez, London, 1966, pp. 195f, 378, 385f, 422, 442, 467, 508, 602.
33. For a penetrating analysis of this kind see Karl Barth, *Church Dogmatics*, II.1, ch. v, where both poles of the epistemic relation are given full discussion. Cf. Barth's statement in *Theology and Church*, London, 1962, p. 342: "Natural theology (*theologia naturalis*) is included in and brought into clear light in the theology of revelation (*theologia revelata*); in the reality of divine grace is included the truth of the divine creation. In this sense it is true that "Grace does not destroy nature but completes it" (*Gratia non tollit naturam sed perficit*). The meaning of the Word of God becomes manifest as it brings into full light the buried and forgotten truth of the creation."

CHAPTER 3

THE SCIENCE OF GOD

THE very title of this chapter will provoke different reactions from people, depending on the notion of "science" fixed in their minds. Some people will find the bringing of "science" and "God" together in this way quite impossible; but this is, I believe, because they operate with a rather rigid and an apparently obsolete conception of what science represents. This means that we shall have to discuss rather fully what science is and what this science is with which we are concerned here.

In the last chapter I argued that although natural theology can no longer be treated as a prior conceptual system on its own, as happened in the ages of cosmological dualism and deism, it must be given its proper place within the embrace of the theology of God's self-revealing interaction with us in the world, for all theology by its very nature can be pursued only within the rational structures of space and time within which we are placed by God, through which he mediates to us knowledge of himself, and within which we may develop and articulate our knowledge of him. I claimed that natural theology must be correlated with this positive theology in much the same way in which we have learned to integrate geometry and physics, when geometry from being an *a priori* deductive system, independent of and antecedent to physics, is transformed into a kind of natural science. By this is not meant, however, that Euclidean geometry is no longer a legitimate pursuit, but that it is to be pursued properly only in recognition that it is an abstract idealisation of real or physical geometry which is four-dimensional in character. The same applies, *mutatis mutandis*, to the new status of natural theology when it is included in and clarified within the real or positive theology of divine revelation.

In order to show the significance of this integration between natural and positive theology let us move over from geometry to logic. Like Euclidean geometry our traditional logic, particularly in its Aristotelian form of class inclusion and exclusion, is an idealisation of formal relations which we prescind from the real world and develop into a formal system of rules for the construction of valid forms of argument. We do this deliberately in order to work out as rigorously and consistently as we can the precise implications of our propositions in reasoned discourse. In this operation, however, we detach formal logic from its ontological correlation with empirical reality, for it is the formal and not the material relations we have in mind. The more we refine our handling of these severely formal relations, which we now do through symbolic and mathematical logics, the further removed they become from experience. The difficulty we have with all formalisations of this kind is that we easily become trapped in them. That is why someone who spends all his life in logic, or in some kind of linguistic, analytic philosophy, is apt to suffer, as it were, from "tunnel vision" with the fixations and hazards that it brings. Now this has undoubtedly been one of the recurring illnesses of theology, most manifest in the great eras of scholastic thought, in mediaeval or in Protestant eras, but it derives largely from the prescriptive role assumed by an independent natural theology which has developed by way of abstraction from sense experience and logical formalisation of conceptual relations. All this *must* change, however, as soon as there takes place a transition from a dualist to a unitary way of thinking, which calls for the integration of natural and positive theology within one bipolar structure of knowledge. The bringing of these two together in this way, the knitting together of epistemological structure and material content, yields what we are bound to call "theological science", that is, not a *scientia formalis* or formal science in the scholastic sense, but a *scientia realis* or real science, the science of

God, and of God in his interaction with the world of space and time.

We evidently have a similar problem today in the development and formulation of quantum theory, where we are unable to construe the kind of connection that comes to view in the behaviour of quanta in terms of Newtonian mechanics or the kind of logic that was necessarily bound up with it.[1] The tension here is so great that the late Friedrich Waissmann used to point out that if traditional logic is right quantum theory must be wrong, and if quantum theory is right, in particular the so-called "uncertainty relation", traditional logic must be wrong, in particular the "law of the excluded middle" which asserts that either a proposition or its negation is true. That is why quantum theory has been forcing some of our theoretical physicists and mathematicians to develop what they speak of as a "three-valued" quantum logic to cope with the distinctive connections between the geometrical and the dynamical aspects of the real world and to help them resolve the paradoxes that arise in formulating quantum-mechanical relations.[2] Until that is done quantum theory cannot be fully or adequately developed. Biologists have found themselves grappling with a similar problem in the need to work out what W. M. Elsasser has called a logic of inhomogeneity as opposed to classical logic which is a logic of homogeneous relations.[3] Modal logic does not really carry us very far in meeting the demands of physicists and biologists in these respects. What they need in their various fields and in different ways is a logic of space-time connections appropriate in field-theory concerned with indivisible continuous dynamic relations, that is to say, something like a "logic of kinetics" which would be a logical counterpart to four-dimensional geometry.

Now inevitably when science takes that immense step forward the whole nature of science changes, for the kind of connection with which it functions at these profound levels, which are intimately correlated with the real world of space and time and organic being, changes. And whatever else science is it is the science of relations in which conceptual relations are developed as expressions of real relations in the universe.[4] Today we live and think in just such an era in which the basic nature and function of our science is in process of radical change, and of course the very concept of science changes at the same time. It is then in this exciting context that I believe theology must be pursued today, as itself the science of the living God. It is a positive and progressive inquiry under the determination of what God discloses to us of his own nature and activity, and within the whole manifold of relations in which we find ourselves as we engage in scientific exploration of the created universe around us.

Let me return to the point I made earlier about tunnel vision. Theology cannot be restricted to the relationship between God and man alone, nor can it be constrained within the narrow limits that are assumed in the regular formalisation of that relationship. Theology has to do with the unlimited reality of God in his relations with the universe of all time and space, and also with man, not merely in respect of his intra-human relations, but in his relations with all that God has made and unceasingly sustains within the universe, for man and the non-human creation are not what they are within the harmony of the universe without one another. An authentic theology will not allow man to be obsessed with himself. Hence we must work hard to break out of our tunnel vision, until we can look around us in the wide world in which God has placed us, and learn to think and speak of God again in that vast *theatrum gloriae*, as Calvin used to call the creation. This is why I have tried to emphasise throughout that we cannot think and speak adequately or worthily of God apart from the whole range of scientific inquiry to which the God-given intelligibilities of the created world are increasingly disclosed to us, for *we* cannot think and speak of God from some place outside the world where God has placed us and where alone he makes himself known to us, and because the *God* whom we seek to know is this God who created the world and is the Lord of all things visible as well as invisible.

It is incumbent upon us, then, as theologians to look about us and to get our proper orientation in the world, the real world, that is, as it comes to view in our various sciences. Now this itself is one of the important doctrines of theology, which I like to express by speaking of man as the priest of creation. Theologically understood man and the universe belong together and together form what we mean by *world* in its relation to God. Man is an essential constituent of the creation, its "crown", as traditional theology has spoken of him, the priest of nature through whose scientific activity under God the inherent intelligibility of the universe comes to expression and articulation. Just as God made life to reproduce itself, so he has made the universe to express itself, to bring forth its own structure and order in ever richer forms, and in that way to find its fulfilment as the creation of God. This is what takes place through man, for man is that unique element in the creation through which the universe knows itself and unfolds its inner rationality. Man is thus by no means a stranger in the cosmos, and his scientific activity is or should not be an imposition *ab extra* upon nature. It is understandable, therefore, that those processes of human thought and activity which have tended to alienate man from nature have led into the chaos of meaninglessness and ecological confusion, for in them man has betrayed his priesthood and misused his creative role in relation to what God has made and entrusted to his responsible investigation. Properly regarded and pursued, scientific activity is not a tormenting of nature but rather the way in which nature pregnant with new forms of being comes to be in travail and to give birth to structured realities out of itself. Man is here the midwife, as it were, and yet rather more than that, for his own rational nature is profoundly geared into the intrinsic rationalities of nature in such a way that he is the appointed instrument under God through which the intelligible universe reveals itself and unfolds out of its crysalis, so to speak, in rational, orderly and beautiful patterns of being. Hence there is disclosed through scientific activity an intelligibility in the created universe beyond man's artifice and control, something absolutely given and transcendent, to which as man he is and must be rationally and responsibly open. That openness and responsibility are part of his human nature as rational agent. Man acts rationally only under the compulsion of reality and its intrinsic order, but it is man's function to bring nature to word, to articulate its dumb rationality in all its latent wonder and beauty and thus to lead the creation in its praise and glorification of God the Creator. That is, as I have called it, the priestly function of man to the creation, within which scientific inquiry becomes an authentically religious duty in man's relationship with God.

So far as theological science is concerned it is imperative that we operate with a *triadic relation* between God, man and world, or God, world and man: for it is this world unfolding its mysteries to our scientific questioning which is the medium of God's revelation and of man's responsible knowledge of him. This implies, however, once more, that there is a necessary and inescapable connection between theological concepts and physical concepts, spiritual and natural concepts, positive and natural theology, or rather between theological science and natural science, for it is in that connection that the changed status of natural theology has its place.

At this juncture we must face certain problems that are unavoidably thrust at us. Both natural and theological science have to start by using common concepts formed in pre-scientific and pre-theological experience and awareness, so that we require to refine these concepts if they are to be used at higher levels of thought and speech and higher levels of communication-systems. However, they must still be coordinated with our everyday experience in this world, and we have to find out how that is properly to be done. If it cannot be done, of course, the concepts we employ are useless for science or for theology. Again theological concepts and natural scientific concepts are related to the universe in different ways, in accordance with the emphasis in each science, in natural science upon looking *at* the universe and its natural order, in theological science upon looking *through* the rational structures of the universe to the Creator. Hence we must be extremely careful in theological science not to read the creaturely content and imagery latent in our this-worldly concepts into the reality they are used to indicate. Theological concepts indicate but do not exhaust or describe the reality to which they refer, but since they refer from our world to God, their worldly starting point is essential to them. How do we make that movement from the world or from man to God in a way that is appropriate to God and yet does not cut off our starting point, for that would make any reference irrelevant for us in this world?

These problems are not peculiar to theology. In mathematics, for example, as Gottlob Frege used to point out,[5] the necessary coordination between number-language and word-language constantly poses problems of this kind. We cannot engage in mathematics simply by using empty symbols, but must employ some other language like German, French or English at the same time if only in order to interpret the symbols and maintain their connection with the real world. In this event, however, we must take care not to read the mental images latent in our word-language into our mathematics and the non-mathematical realities upon which its equations bear. We have a similar problem in relativity and in quantum theory where we must be careful not to project our visual images into the inherently invisible space-time metrical field or into the world of sub-atomic particles.

Problems of this kind are quite basic. They impose on us the task of working out the *ana-logical reference* of our concepts and terms whether in natural or in theological science. I am not thinking of "analogy" here as the great mediaeval thinkers, for the most part, employed it, in terms of proportional relationships transferred from Greek mathematics. I am thinking rather of analogy as the tracing of our concepts back to the source which gave rise to them. This is not a logical movement of thought on one and the same logical level, but a movement of thought across logical levels. It represents the reading back (the *analegein*) of our concepts from one level to another or higher level in such a way that we do not impose upon the reality indicated on that level the images and ideas we derive from a lower level, but rather in such a way that we allow our concepts to be controlled by the reality they intend beyond their own level.

Now while knowledge of God involves a two-way relation between our knowledge of God and our knowledge of ourselves, it is concerned to pierce through that reciprocity to knowledge of God in his own Logos and Being. It is in that way and in that light that theological concepts are formed and refined out of pre-theological concepts purified of their irrelevant overtones or culture-conditioned deviations, but without resolving away their empirical correlation with the spatio-temporal structures within which human knowledge of God is always actualised. On the other hand, our natural scientific knowledge which aims at understanding the universe in terms of its intrinsic intelligibilities is also actualised within the same spatio-temporal structures, and its concepts are gained and refined from pre-scientific concepts, without loss of empirical connection with those structures. This overlap in our knowledge of God and our knowledge of the universe, and therefore in the empirical correlates of theological and scientific concepts, means that we must attend to the differences between natural science and theological science, the way of understanding the world by looking *at* it and the way of understanding the world by looking *through* it. What is distinctive of theological science in this respect is the blend it involves between scientific inquiry and the worship of God in which our human thought is heuristically elevated to higher levels of reality beyond what is merely palpable, tangible and visible. This is a point to which we shall have to return in this chapter but to which we must give rather more attention in a later chapter in which we will be concerned with the social coefficient of theological knowledge.

Let us now return to the question we raised earlier: What

is science? A gradual but profound change in the concept of science has been taking place ever since the introduction of concepts of the continuous field through Michael Faraday and James Clerk Maxwell, and especially since the immense impetus given to field theory by Albert Einstein.[6] Scientists are undoubtedly still struggling with this change, but it is every day more evident that the concept of science is shedding its older hard, mechanistic, instrumentalist character, and taking on a more subtle, elastic form appropriate to fields of dynamic and organic relations. As this becomes more and more the case sociologists will not be so tempted to reduce personal and social relations to the kind of connections that we used to operate within classical physics and chemistry. Historians will likewise cease to yield to the same kind of temptation in the way they handle evidence and the kind of criteria they use in determining connections. In other words, the variational and differential principles which natural science has had to deploy more and more since Newton are now affecting the fundamental concept and shape of science in every sphere of research. It is science of this kind that is increasingly demanded of us as the extraordinarily rich and manifold nature of order in the universe comes increasingly to view, for the narrow and rather inflexible concept of science fails again and again to match the actual modes of connection found to exist in nature, and is in fact proving an obstacle to scientific progress. This has long been evident in biological science, but is no less evident, as we are well aware, in several of the social sciences, including psychology and history.

There is another factor that must be taken into account in the reappraisal of science. This has to do with the old Cartesian split between subject and object, and the setting up as the prime model of knowledge that of the observing and thinking subject in a detached, impersonal relation to the object. This lies at the root of the problem which we have already had occasion to discuss once or twice in different contexts, in the transition of the modern mind from an inherent to a technological rationality, together with the deep cleavage between understanding (*Verstehen*) and explaining (*Erklären*) which came to the front in the nineteenth century and has left its mark in twentieth-century thought. Science is not really determined by the cognitional structure of the rational agent, yet it does function with an astonishing affinity between the rationality of the subject and that of the object. This is evident not only at the very start of research when subject and object are posited together in our actual experience and its pre-scientific interpretation, but in those great scientific achievements where profound change in the theoretic basis and structure of knowledge takes place, and the laws of nature and the laws of thought are found to harmonise with each other.

There is certainly no way to explain physical features in the universe by reference to structures of the human understanding, for the understanding that seeks to explain things beyond itself in terms of relations and patterns within itself inevitably goes wrong. It is more and more apparent that the proper nature of the rational subject and the true functioning of his understanding are revealed only as we break through the subject-object bifurcation and allow our minds to function as they must under the constraint of objective reality and the compelling claims of its inherent intelligibility. Thus it is not only the rationality of the universe that comes to view in our scientific operations, but the rationality of man himself, but that has a powerful feed-back effect upon the reconstruction of our conception of the scientific enterprise.

If Einstein was correct, as I believe he was, the principal obstacle to progress in appreciating the real nature of scientific activity was the widespread idea that scientific concepts are derived by way of logical inference from sense experience.[7] That did not represent the way in which Newton had actually operated in developing his system of the world, but it was an interpretation of his method which he himself sometimes gave, and which was exploited by the rising tide of rationalist empiricism in its exaltation of the autonomous human reason. When we proceed by the abstraction and deduction of concepts from observations under the guidance of the Aristotelian principle that there is nothing in the mind which was not first in the senses, we soon find that we cannot account in this empiricist way for certain fundamental notions such as causality.

In that event we need something like a Kantian critique of pure reason within the framework of the Newtonian system in order to provide us with *stability*, with a consistent structure of intelligibility, on which we can reply in all our scientific and rational operations, if only because abstractive procedures lend themselves to the creative and arbitrary and undependable spontaneities of the human mind. Kant's critical philosophy provided the scientific reason with in-built forms of intuition and categories of the understanding bearing upon substance and causality, time and space, of an "absolute" kind, that is, unaffected by experience, and put them forward as the *regulative* structures within which rational account could be given for Newton's laws of motion. Thus in spite of empiricist criticism he was able to provide on-going science with a *working objectivity*, but at the expense of any hope of being able to apprehend things in themselves in their interior relations independent of the way they appear to us and of our observations of them. Kant was certainly right in insisting that scientific concepts have content only when they are connected with observational experience, but his concept of the synthetic *a priori* had the effect of limiting that connection to the phenomenal surface of experience, and thus seriously obstructed development through the inertia or immobility of his operative framework. That way of thinking, of course, has been shattered through the imperious demands of empirical fact for a more satisfactory theoretic basis.

As we look back upon observationalist and abstractive science of this kind, it is not difficult to discern its immense drawbacks. It built into the structure of science basically idealist preconceptions which have been very difficult to eradicate and which keep on obstructing advance in scientific knowledge at decisive points. I think, for ex-

THE SCIENCE OF GOD

ample, of the twenty years' struggle with idealist notions of space and time which Einstein had before he could arrive at relativity theory, and the difficulties which we still have with quantum theory, particularly as it stems from Bohr, Heisenberg and Born, which may be traced, in part at least, to Kantian presuppositions. Underneath all this there lurk the persistent problems of phenomenalism and dualism, relating to notions of sense data and logical deduction from them but not least to the prescriptive character of observationalist and empiricist epistemology, whether in its operationalist or conventionalist forms. Where is the effect of this outlook more evident than in the so-called "Vienna Circle" with its logical positivism and verificationism which hang together with this phenomenalist and observationalist approach to science?[8]

It must also be added that abstractive procedures of thought have had the effect of hardening the Newtonian framework of science into a rigidly determinist and mechanistic system. And so from Laplace to Haekel we find the tyrannical dominance, especially in Continental science and philosophy, of the closed mechanistic universe. Our recent science has found, however, that it could only advance more deeply into the understanding of empirical realities when it broke free from the artificial limitations of determinist and mechanistic concepts. This is especially evident in various developments of field theory as repeated attempts to interpret the Faraday-Clerk Maxwell concept of the electromagnetic field in terms of Newtonian mechanics broke down again and again. Broadly speaking, science is still recovering from the shock of that failure and learning slowly to adjust itself to the more fluid and dynamic notion of continuous fields of force.

I find that it is when an adjustment of this kind has not been made, that people find it difficult to think of a science of God, for in their minds this seems to mean that we are trying to manipulate God through objectifying forms of thought and even to impound him, as it were, within the mechanistic universe with its rigid system of a closed continuum of cause and effect, whereas they wish, in assuming the finality of such a mechanistic and causalist outlook upon the universe, to relegate God to some utterly transcendentalist realm beyond any interaction with our world of space and time. However, where it is clearly recognised that science by its very nature must operate with modes of rationality prescribed by the intrinsic intelligibility of that into which it inquires, and develops corresponding modes of discovery, interpretation, articulation and verification, then once again real cooperation between natural and theological science becomes not only possible but imperative, especially when it is also recognised that theological science is pursued only within the same structures of space and time within which natural science operates.

The fundamental difficulty with abstractive and positivist science, as I have already argued, is that it operates with a logical bridge between concepts and experience, both at the start and at the finish, that is, in the derivation of concepts from the universe as we experience it and in the verificatory procedures relating concepts back to experience.[9] This way of thinking operates with a set of fixed *principia* which are either unquestioningly assumed or held to be definitely verifiable and which if not originally *a priori* at least play an *a priori* role in the development of knowledge. Now this is not only a difficulty but an impossibility, for there is not and cannot be any logical bridge between ideas and existence.[10] There is indeed a deep and wonderful correlation between concepts and experience, and science operates with that correlation everywhere, but since there is no logical bridge the scientist does not work with rules for inductive procedures, and cannot finally verify his claims to have discovered the structures of reality by logical means. This does not imply that there are not immensely important logico-deductive processes that have to be undertaken in the construction of scientific theories and in testing their consistency, but the actual way in which they are applied to empirical existence, which is such a crucial test, is not basically different from the way in which they are discovered by the scientist in the first place.

Einstein himself has perhaps done more than any other to put an end to the abstractivist notion of science, not just by what he has argued but by his deeds of achievement in advancing our knowledge of the universe in an unparalleled way in which he broke free from the early influence on him of Ernst Mach, the "high priest" of positivism as he might be called.[11] Yet according to Einstein "nothing can be said concerning the manner in which concepts are to be made and connected, and how we are to coordinate them to the experiences".[12] Nevertheless Einstein does give us some indication of how he himself operated. In this connection he used to stress the role of intuitive apprehension in the scientist's grasp of reality.[13] He also described the way in which he developed his ideas as "free creations" of his thought.[14] By that he did not mean that they were mere fictions or empty fantasies, but that they arose prior to his reasoning processes and had a free, spontaneous character. They did not arise out of abstractions from observations, nor were they derived according to logical rules. They arose out of an intimacy with and a sympathetic understanding of experience, under the belief in the intelligibility or comprehensibility (*Verständlichkeit*) of the world external to the percipient. The fundamental convictions with which he worked were freely chosen but they were nevertheless controlled through his intuitive apprehension of the structure of reality itself, and were confirmed directly or indirectly by their applicability to empirical reality.[15]

Michael Polanyi has spoken of this Einsteinian process of thought as "a new 'epistemological' method of speculative discovery".[16] Through his enlightening analyses and exposition in book after book he has gone farther than any other scientist in making this process of inquiry and theoretic construction as clear as possible, and in his own delicate contributions he has shown us that creative scientific discovery of this kind is *unformalisable*.[17]

Polanyi and Einstein are here in the deepest agreement

in regard to the way in which concepts and experience are coordinated, and in the postulation of unspecifiable clues as free axioms which, through the creative insights of the scientist, are formed into a heuristic instrument for axiomatic penetration into the objective structures of the field of reality being investigated. Many people are rather puzzled by this use of "axioms", but let me stress that these axioms are known only in and through the on-going processes of interrogation and discovery, and not beforehand, except perhaps in some tenuous subsidiary way. We are not concerned here with axioms in the classical sense of fixed premisses or *principia*, when we argue from certain *a priori* or accepted premisses to conclusions, for example, in Aristotelian science or in Euclidean geometry. Those axioms or *principia* are closed at both ends, for they fix the starting point and determine the conclusion. But what we are concerned with here are open flexible concepts used in a fluid axiomatic way, and therefore with constant revision the further they penetrate into and lay bare the inner logic of the chosen field of investigation.

This axiomatic penetration into the mysteries of nature involves the development of appropriate conceptual and symbolic forms as instruments by which reality may be "grasped in depth" or by means of which the scientist can "worm his way" into the intrinsic features and properties of reality.[18] Polanyi has shown that where Einstein speaks of intuitive insight or apprehension we operate with an *anticipatory* grasp or a preliminary inkling of the rational pattern of things which is essentially an inarticulate movement of thought or a tacit form of apprehension. It is through reliance upon this implicit recognition of how things really are that our scientific inquiries progress, when explicit and more specifiable modes of conception and articulation arise leading to the formation of definite theories. Yet in the application of these theories to existence we rely once again on unformalisable acts of judgment, for here as at the beginning we have to do with the relation between our concepts and empirical reality which cannot, in the nature of the case, be resolved into the relations of ideas with one another or therefore be given explicit formulation.[19]

Undoubtedly an essential and crucial factor in the advance of scientific inquiry, especially at decisive points of radical change and theoretic reconstruction of the basis of knowledge, is the cultivation of what Clerk Maxwell called a *new mathesis*, through which we allow fresh revelations of nature to call forth from our minds new modes of thought by which the process of our minds can be brought into the most complete harmony with the process of nature.[20] What is needed is the discovery and formation of apposite conceptual instruments which we may use in breaking through to the deeper layers of the intelligibility immanent in the universe and in reshaping our own understanding and outlook upon the universe in such a way that they are matched to it. Einstein used to point to Newton and Clerk Maxwell as exhibiting fundamental shifts of this kind in our understanding of the universe and in the axiomatic substructure of physics.[21] Thus in grasping and working out an understanding of the universe as a vast system of bodies in motion, Newton devised the concept of "fluxions" or differential quotients by means of which he established the differential laws of nature and laid the foundations of classical physics and mechanics. But with Clerk Maxwell another radical shift had to be undertaken when he devised his partial differential equations through which a thoroughly relational and dynamic understanding of nature in terms of continuous fields of force could be its natural and rigorous expression but through which at the same time the fundamental ideas and convictions governing scientific inquiry and theory began to be changed. Thus with Clerk Maxwell an immense advance was made in understanding the real patterns of relations immanent in nature which had to be expressed by a new type of physical law — that is what he achieved through his partial differential equations, but at the same time those equations exhibited themselves as "the natural expression of the primary realities of physics".[22] To Einstein's examples, of course, we have to add that of

Einstein's own and no less radical transformation of our understanding of the universe and of the fundamental structure of science that came with the special and general theories of relativity.

When in this light we look at the whole development of western natural science, we might well claim that there have been three periods of major advance and radical shift in our understanding of science, which correspond in their way to the three great eras of change in cosmological outlook. First, the early classical development, from Pythagorean and Ptolemaic times right up to the Newtonian era, which operated with deductive geometric conceptions. Second, the Newtonian era of causalist science concerned with the particulate view of nature, which operated with differential conceptions and variational principles in offering an account of a universe of bodies in motion. Third, the modern era of field theory in which a new science, rejecting the dualist bases of the two previous eras, understands nature in terms of continuous indivisible fields in a multilevelled universe, and operates with partial differential equations, four-dimensional geometries, mathematical invariances, quantum-mechanical structures, and so on.

At each of these stages we have taken a great stride into a deeper understanding of created reality. But today it is very apparent, in the rapid acceleration and sophistication of scientific inquiry, that we are ourselves in an age where new modes of thought, new cognitive instruments, and new material logics, are required, if we really are to think out and express the rational connections between the dynamical and geometrical aspects of reality (such as come to light in quantum theory, for example) in a way that is genuinely appropriate to them and that at the same time brings the processes of our minds into harmony with them. Certainly the more deeply we grasp the inherent intelligibilities of the universe, the more it surprises us by its revelations, and the more it evokes from us wonder at the untold possibilities of further unexpected disclosure, calling for ever new modes of thought and conceptual instrumentation far surpassing the scientific machinery of the past.

Now I believe very firmly that it is within this unfolding of the created universe to scientific inquiry and under the compelling claims of its God-given intelligibility that we theologians have to engage in our own distinctive inquiries. And there can be little doubt that theology in its own appropriate way must operate along similar lines — that is, not in accordance with abstractive processes of thought, nor with argumentation from fixed principles, but rather in accordance with the fluid axiomatic modes of thought that have served so well to reveal the mysteries of nature. In this case, theology must take great care to develop its own proper axioms under the pressure of God's revealing interaction with us in space and time, i.e. fluid axioms subject to constant revision in the light of the divine revelation which they serve and seek to bring to articulate and coherent understanding in our minds. The rejection of abstractive procedures and ways of reasoning from fixed axioms applies to the old-fashioned operations of natural theology which were, as far as I can see, essentially of this kind. Authentic theology cannot be built up in this way.

This applies also, as far as I can see, to the ways in which historico-critical methods have been applied, particularly by Protestant scholars, to the understanding and interpretation of the Holy Scriptures, within a general framework that is still governed by dualist and phenomenalist assumptions which do not admit of knowledge of things in themselves or in their own intrinsic significance but only as they appear to us. This is nowhere more evident than in the investigation of the historical Jesus that has been undertaken for at least a century and a half, and in the handling of the New Testament accounts of Jesus, where observationalist and abstractive processes have been at work with the effect of disrupting the original and natural integration of empirical and theoretical, historical and theological, factors in the kerygmatic and didactic presentation of the Gospel. Then with the fragmentation and disintegration of the evangelical source material as a result of these abstractive and analytical methods, scholars keep searching for some other kind of framework in which at least some of the appearances may be saved and fitted into a coherent pattern, but that inevitably means the imposition of an alien and artificially contrived pattern upon their "findings" or "scientifically established facts". Then processed "data" of this kind are handed over to the theologian for his expected logico-deductive operations, but they prove to be seriously mutilated and misleading for they are shorn of their intrinsic significance. It is certainly not along these lines either that an authentic and rigorous theological science can be pursued, in its commitment to the inherence of theoretical and empirical ingredients in primordial experience and the foundations of knowledge, and therefore to the inseparability of form and content or method and subject-matter in its analytical or constructive operations.

How then do we proceed in the science of God?

Formally, the procedure is the same as that which we find in the functioning of natural science, but materially there are significant differences that have to be taken into account, differences that are scientifically required through fidelity to the nature of the reality into which we inquire and the features of the field in which we have to operate. This implies the painful experience of what the mystics call "unknowing" or "unlearning", for we have to divest our minds of alien and unwarranted presuppositions and also of what we claimed to know beforehand. The proper nature of the reality and the distinctive features of the field become theologically clear to us only as we proceed, but as and insofar as they become clear we must learn how to revise our fundamental convictions and axioms and how to devise appropriate modes of thought and relevant cognitive instruments to match the intrinsic intelligibilities that come to light.

The supreme difference of course lies in the fact that the reality with which we have to do in theology is God himself, the Creator of the universe, who is active Agent and not just a detached and inert deistic postulate. The active creative nature of the divine Agent, and the actual form which his self-revelation and self-communication to mankind has taken, in the incarnation of his Word and Love in Jesus Christ within the contingent structures of space and time, mean that the nature of the corresponding community in which theological inquiry takes place is also very different. It is not just the community of verifiers with which we operate in the functioning of natural science — it is perhaps that also but not just that — but the community of believers living in empirical contact and communion with God, who allow their minds and lives to take shape under the impact of what God reveals of his own nature and truth.

Theological science must therefore generate concepts of God that are *worthy* of him, as the great Origen used to insist,[23] not concepts which we devise and project out of our own religious self-understanding but concepts that are forced on our minds by the sheer nature of the divine Majesty. And that can take place only through intimate experience of God within the communion of people with God that arises in the world under the impact and imprint of his revealing and redemptive acts. This is not a spaceless and timeless community but one existing and taking shape within the space-time structures of the world, and within the orbit of heuristic worship and openness of being to which we have already alluded in the correlation of theological and natural science. We are concerned here not only with the reciprocal relations between man and God which he calls into being but with those relations as they are grounded and ordered within physical and historical existence which is the created medium of rational communication between God and man. In other words, theological science functions within the Church regarded as the spatio-temporal community of reciprocity which God establishes in his self-giving to mankind. We shall return to this theme in the following chapter when we come to discuss the social coefficient of the knowledge of God.

The first step to be taken here, then, is rather similar to that of which Einstein and Polanyi have spoken in their own spheres of scientific inquiry, namely, an empirico-

84   REALITY AND SCIENTIFIC THEOLOGY

[STEP 1] intuitive movement of thought in which we cultivate incipient insights into the objective patterns or configuration of our chosen field. These are what we call *clues*, informal glimpses of reality, pointers to reality, or aspects of reality pressing for recognition in our minds. As we have seen, they are essentially of an anticipatory nature, anticipatory because they come from reality and draw us toward it. There are no formal rules for acquiring these enlightening intimations of reality. Only the great theologians who were also childlike in spirit have been able to come up with the basic insights and fundamental ideas that have advanced theological understanding.

This sort of clue, of which I have been speaking, is what Clement of Alexandria (who preceded Origen as head of the Catechetical School in Alexandria in the last decade of the second century) called a *prolēpsis*, a forward leap of the awakened mind in laying hold of some aspect of reality, not a preconception or a presumption formed on the basis of prior knowledge but the first step in conceptual assent (*ennoetikē sygkatathesis/pistis*) to a hitherto unknown truth which can be known only through the force of its own self-evidence. It is under the guidance of a proleptic experience or apprehension of God in this way that we may engage in an authentically heuristic process of inquiry and knowledge (*heuretikē ekzetēsis/epistēmē*) leading through scientific perception (*epistēmonikē theoria*) to *accurate* faith (*akribē pistis*), i.e. a rational assent that corresponds precisely to the nature of the reality apprehended. In the prosecution of such an open yet disciplined inquiry we move beyond our false preconceptions and begin to frame the main cognitive instruments we need to build up knowledge of God in his own Being and Truth.[24]

If it is asked where and how such a *prolēpsis* arises, the answer must be, within the community of the worshipping people of God and their evangelical fellowship with one another in the service of the Gospel. It takes shape in our minds as we meditate upon the Holy Scriptures, listen together to God's voice addressing us through them, and let the Word of God dwell in us in such a way that it becomes interiorised in our understanding.[25] It is as we live and think together in this way, in accordance with godliness (*kat'eusebeian*),[26] as Clement, Origen and Athanasius, all used to insist, that under the imprint of God's Holiness and Love there arise in our minds modes of thought that are worthy of him. Through the creative power of his Word and Spirit we are enabled to know God in some real measure as he is in himself and in a way that would not be possible otherwise.

Let it be granted, then, that the basic convictions and fundamental ideas with which our knowledge of God is built up arise on the ground of evangelical and liturgical experience in the life of the Church, in response to the way God has actually taken in making himself known to mankind through historical dialogue with Israel and the Incarnation of his Son in Jesus Christ and continues to reveal himself to us through the Holy Scriptures. Scientific theology or theological science, strictly speaking, can never be more than a refinement and extension of the knowledge informed by those basic convictions and fundamental ideas, and it would be both empty of material content and empirically irrelevant if it were cut adrift from them. Thus in any development of theological science we must take care at all levels to maintain intuitive contact with immediate experience, knowledge and worship of God, for that determines its cognitive value.

Its specific task, however, as a science is to clarify and unify knowledge of God in such a way that we can bring some kind of theological order into our everyday experience and faith in God and thereby to promote further and deeper understanding of him. Hence in theological science we select a few primary convictions and ideas as close as possible to the ground of experience and faith and organise them into a preliminary "model" through which we seek to penetrate into the essential configuration of the field of inquiry in order to trace something of its inner connections. That is to say, we devise and deploy a revisable, flexible axiomatic instrument in determining the rational structure of theological knowledge and in bringing to light its inner logic. If the concepts we initially select prove to be inappropriate and fail to fulfil our expectations, we try again with another set, or another, which we judge in the light of the experience we gain to be more promising, until we find a set that brings results in enabling us to probe deeply into the configuration of the field in such a way as to let us discern something of its orderly cohesion and pattern. Then under the guidance of what we learn in this way we proceed to construct higher-level disclosure models through which our grasp of the inner logic of divine revelation becomes clearer and more explicit, so that the way is opened up for general clarification and simplification of the whole body of theological knowledge. Moreover in its feed-back application to the fundamental ground of our experience and faith such a clarification and simplification will have the effect of allowing the informational content of God's self-revelation to bear upon our understanding at basic theoretical points, or doctrines, with greater precision and force.

It may help at this juncture to offer some examples from the history of Christian theology of this sort of axiomatic procedure in developing our understanding of the knowledge of God.

Let me refer first to Athanasius of Alexandria who in the fourth century more than any other set Christian theology on its scientific basis. I have in mind his twin monographs, *Contra Gentes* and *De Incarnatione*[27] which together provide us with a clear example of the kind of rational activity we employ as an attempt is made through heuristic argumentation within the field set up by God's self-revealing interaction with man in the world to find a way into the central order of things which is then allowed to throw light upon the whole complex of connections with which we have to do in theology.[28] All *a priori* arguments are set aside and any argumentation from an epistemological or cosmological system people may have inherited prior to or independently of their actual knowledge of God as the Father of Jesus Christ. Nor is there any attempt made to derive knowledge of God abstractively from the Holy Scriptures or out of the manuals of earlier theologians, but rather through a reasoned movement of thought within the field of Christian experience and faith to penetrate into its intrinsic order and intelligibility.[29] The *Contra Gentes* and the *De Incarnatione* have thus great methodological significance in the history of theology, for they broke new ground and put forward a new scientific method in showing how a conjunctive and synthetic mode of thought could probe into the intrinsic subject-matter of theology with fruitful results: in disclosing the organic way in which creation and redemption are to be understood from a point of central reference in the Incarnation of the Word or Son of God, and in developing an intelligible structure of understanding reaching back to a creative centre in God himself, which throws an integrating light upon all theological relations and connections.[30]

In regard to the *Contra Gentes*, let me simply remark that it cannot be construed as providing a natural theology in the traditional sense, which as we have seen has always arisen out of a dualist framework of thought, for that kind of framework is explicitly rejected by Athanasius who does not operate with a distinction between the supernatural and the natural that the disjunction between intelligible and sensible realms implied.[31] On the contrary in the *Contra Gentes* we are presented with a proper natural theology as the polar complement to revealed theology which is the subject of the *De Incarnatione*, and which is revealed to have its own inherent reasonableness precisely through sharing in the same reasonableness manifest in the Incarnation of the divine Logos or Intelligibility in the Incarnation.

In the *De Incarnatione* Athanasius turned his attention explicitly to the relationship between God and man revealed in the person, word and work of Jesus Christ. Through probing questions one after the other he set aside unsuitable, or unfitting or inappropriate notions in order to find what was really in accordance with the nature of God and man in their interrelations, in a sustained attempt to get hold of the appropriate kind of connection in thought which will match the inner sequence (*akolouthia*) of divine revelation or the chain of reason (*heirmos*) manifest in God's loving action toward mankind, the divine *philanthrōpia*, the logic of the Love of God become man. That is to say, Athanasius set himself to the disciplined task of uncovering and determining the ordering force and distinctive pattern of grace (*hē kat'eikona charis*) revealed in the incarnate Logos in order to learn from the Logos himself, that is, from the intrinsic Intelligibility of God hypostatically embodied in him, as far as it is possible for the human understanding to grasp it, the inner movement and reason for God's redeeming and renewing activity in the cosmos.[31]

Now let us jump over the centuries and take our second example of a fluid axiomatic movement of thought from St. Anselm of Canterbury in the eleventh century, and specifically from his work on the Incarnation known as *Cur Deus Homo*.[32] Here again we find a way of open inquiry that refuses to operate logico-deductively from fixed *principia* or traditional authorities, whether they are ecclesiastical or biblical, but insists on keeping close to the ground of actual faith and experience.[33] In recognition of the fact that faith itself does not rest on biblical, far less on ecclesiastical, authority as such but on the truth mediated through the Bible and the Church, Anselm proposed a way of inquiry which methodologically sets aside even biblical statements regarded as formal premises, or which passes through them to the solid truth (*solida veritas*) on which they rest, in order that the mind may be brought directly under the compulsion of the truth and the impress of its rationality.[34] Even in Christology itself Anselm declined to treat Christ as a formal premiss or a propositional base for logical operations, but setting him aside in that role (*remoto Christo*, or *quasi nihil sciatur de Christo*), and with constant prayer for divine illumination, he found a way of probing into the heart of Christological knowledge and elucidating its inner logic so that faith in Christ and knowledge of God through him could be shown to rest directly on the rationality of the truth (*ratio veritatis*) incarnate in Christ.[35]

Quite clearly Anselm was not impugning the authority of the Holy Scriptures or of the doctrines of the Faith, but was establishing them all the more firmly on their proper foundation, upon the reality of things as they actually are and as they ought to be known and expressed by us.[36] He worked with a hierarchy of different levels in which his thought moved from the truth of statement through the truth of being to the Supreme Truth of God.[37] This had the effect of clarifying the coherent structure of theology and of showing that theological concepts are formed and theological statements are made *rightly* only when they point beyond themselves to the Truth of God to which they are indebted as their Source. But it also has the effect of showing that, while all theological concepts and statements are inadequate, for God infinitely transcends all our thought and speech of him, nevertheless they are not for that reason necessarily false, for their truth as concepts and statements does not rest in themselves but in him to whom they refer. Expressed the other way round, Anselm showed that since God makes his own supreme Truth the objective ground of our knowledge of him he thereby confers relativity upon it. Thus theological inquiry and humility go hand in hand.[38]

Now in the third instance let us move to modern times and a very different kind of person, Søren Kierkegaard, and that fascinating document called *Philosophical Fragments*.[39] Once again we have a thinker who rejected the patterns of formal argument and engaged in an open-structured movement of thought which, judged from the perspective of logic, represents no more than a set of fragments. The Incarnation eventually emerged as the real theme, but Kierkegaard came at it from behind and under cover of a kind of theological anonymity.[40] He did not proceed by arguing deductively from the Bible, Revelation, Christ or Church dogma, as fixed principles, but in his characteristically humorous and ironic manner he seized upon one apparently tenuous clue after another and used it to penetrate ingeniously into the very substance of the Faith until the essential structure of the Christian doctrine of the Incarnation and our relation to Christ as disciples was forcefully brought out.

Throughout the work, however, Kierkegaard developed a sub-theme which turned out to have the greatest significance, the relation of the truth to *time*, which had been conspicuously missing from Anselm's thought. Behind Kierkegaard's concern, as we can see from some of his other writings, lay his engagement with problems he found in Aristotle, Kant, Lessing and Hegel and the stimulation of some ideas he derived from Trendelenburg's critique of Kantian notions of time.[41] But what really gripped Kierkegaard and forced him to come to terms with it was the fact that in the Incarnation "absolute" truth moved into time in Jesus Christ and became "historical fact", which implies that we cannot know the truth except in a dynamic way involving a temporal or historical relation to it. If the truth has *moved* into time and *become* historical event, then movement or *kinēsis* belongs to truth and has categorical significance.[42] And so the question had to be raised as to how the reason can apprehend and think movement or kinetic truth, and in particular the Truth of God that has become active, personal being in Jesus Christ, without undergoing a radical transformation in conformity to the truth. In wrestling with this problem of *transition* Kierkegaard found he had to abandon a way of thinking from a point of absolute rest, and opt for a kinetic mode of the reason with which to apprehend movement, continuity, dynamic truth, without resolving them into something quite different in terms of static necessities or timeless possibilities.[43] He referred to his act of the reason variously as a decision, a resolution, or a leap,[44] and spoke of faith as having the required condition.[45]

This is surely one of the really significant moments in the history of thought which was matched by a younger contemporary of Kierkegaard, James Clerk Maxwell when he called for a "dynamical mode of reasoning" or "physical reasoning" as the appropriate way to understand and interpret the electromagnetic field.[46] But in abandoning a way of thinking from a point of absolute rest Kierkegaard took a step in theology much like that which Einstein in the next century had to take in reaching relativity theory on the basis of Clerk Maxwell's concept of the dynamic continuous field of radiation. The significance of Kierkegaard's step is thrown into sharp relief when we consider it from the perspective of logic, as he challenged his readers to do, for traditional logic eliminates the place of action and time in our thought. To cite Kierkegaard himself: "In logic no movement can *come about*, for logic *is*, and everything logical simply is, and this impotence of logic is the transition to the sphere of being where existence and reality appear."[47] The fact that logical operations reduce movements involving time to vanishing points is something that we have yet to think out properly in the modern world, not least in view of the inadequacy of so-called "tense logics", for it will not do in logic to treat time as an external parameter any more than it will do in geometry. It is precisely in breaking through that Newtonian way of thinking that relativity theory and non-equilibrium thermodynamics have had to pioneer new ways of explaining the kind of dynamic order we find in nature.

To return to our immediate interest in this chapter, I believe that theological science must learn from the three great classical thinkers whom I have adduced, and combine their different insights into the distinctive kinds of connection that come to light in theological inquiry, with a view to finding the adequate cognitive tools with which we in our day can penetrate more deeply into the inner structure of the knowledge of God and do full justice to the dynamic interaction of God with mankind in the universe of space and time. What we evidently need is the elaboration of a new material logic, a *theologic*, appropriate to the nature of the Incarnation, which treats spatio-temporal factors as internal parameters in our interpretative and explicatory formalisations of God's self-revelation.

I believe also that Christian theology has not a little to learn in this respect from the rise and development of the field theory of light, from James Clerk Maxwell to Albert Einstein and beyond. Let me explain. In the particulate view of nature as it developed from Newton there arose a corpuscular theory of light, but then with an understanding of nature in terms of moving fields of force there arose an undulatory theory of light. While these theories have proved to be complementary and can be theoretically coordinated, they exhibit a dualism between the material point and the continuous field in the scientific description of the behaviour of light that is unsatisfactory and disturbing.[48] Some kind of unitary theory is required to transcend the contraposition between particulate and undulatory concepts of light. But will that really be possible unless a new kind of logic is developed in which the ontic and dynamic features of light can be brought together? As we have already seen, this is the sort of thing that various people have actually been working on in connection with so-called "quantum logic".

Now it seems evident to me that this is also the sort of task that theologians need to undertake. Classical Christian theology, in late Patristic and Mediaeval times and in the age of Protestant scholasticism, was absorbed with ontic relations and structures, but modern theology stemming from the Reformation has more and more stressed dynamic relations and structures. To a certain extent they may be regarded as complementary movements in the history of thought, but being and event, ontic and dynamic aspects of reality, torn apart become nonsensical, which is a very unsatisfactory state of affairs, although that is unfortunately the situation in which much contemporary theology has become trapped. This surely calls for a thorough reconstruction of Christian thought, in which attention is focussed upon the integration of substance and act, or being and event, at all levels of theological inquiry and formulation, in a way not unlike what is now taking place in natural science, whether in physics or in biology.

It is my purpose, then, to call for a new rigorous theology in the form of *axiomatic dogmatics*. By this I mean dogmatics in the classical form devoted to positive knowledge of God grounded in and informed by divine revelation, but a dogmatics operating with fluid axioms in the modern style instead of with fixed axioms or principles in the static style of the past. It must be emphasised, however, that a proper account of the axiomatic procedure of theological science can be given only along with an interpretative account of the material or doctrinal content of Christian knowledge of God, for that is the theoretic substance of the faith that is to be axiomatised in dogmatic science. On the other hand, it must also be emphasised that since we do not operate here with axiomatisation in the old sense leading to closed logico-deductive systems, we have to reckon with the development of open dogmatic structures. By this is meant dogmatic structures which cannot be complete in themselves if they are to be meaningful and consistent. They must have constituent elements or basic concepts that are not decidable or provable within the systematic organisation of the general body of our theological concepts. That is to say, dogmatic theology may be axiomatised in a consistent and meaningful way only if at decisive points it is correlated with *mystical theology*, for it is in mystical theology that the boundary conditions of our dogmatic formalisations are kept open toward the transcendent and unlimited Rationality and Freedom of the living God. Dogmatic theology of this kind is informed by concepts that are both worthy of the nature of God and empirically relevant in the world he has made.

NOTES

1. When the connections described in classical mechanical terms and the connections expressed in traditional formal logic are thrown into algebraic form they are found to be variants of the same thing.
2. F. Waissmann, *How I See Philosophy*, London, 1968, p. 252. See also L. E. J. Brouwer, in *Philosophy of Mathematics. Selected*

Readings, edit. by P. Benacerraf and H. Putnam, Oxford, 1964, p. 78f; E. T. Whittaker, *Space and Spirit*, Edinburgh, 1964, p. 38f; Max Born, in P. A. Schilpp, *op. cit.*, p. 177; and Hans Reichenbach, *Philosophic Foundations of Quantum Mechanics*, Berkeley and Los Angeles, 1946, pp. 144ff.
3. W. M. Elsasser, *Atom and Organism*, Princeton, 1966, pp. 14, 23ff, 31ff, 42ff, 57, 77ff, 128. Cf. A. L. Lehninger, *Biochemistry*, New York, 1970, p. 7f, on the "molecular logic" of open systems.
4. Cf. James Clerk Maxwell, "In a scientific point of view the *relation* is the most important thing to know." Lewis Campbell and William Garnett, *The Life of James Clerk Maxwell, with a selection of his correspondence and occasional writings*, London, 1882, p. 243.
5. G. Frege, *Die Grundsetzte der Arithmetik*, Jena, 1903, Bd. II. § 91f. Cf. F. Waissmann, *Introduction to Mathematical Thinking*, New York, 1959, pp. 235ff.
6. See J. Agassi, *Faraday as a Natural Philosopher*, Chicago and London, 1971; and W. Berkson, *Fields of Force. The Development of a World View from Faraday to Einstein*, London, 1974; A. Einstein and L. Infeld, *The Evolution of Physics, From Early Concepts to Relativity and Quanta*, New York, 1938, part III.
7. A. Einstein, *The World as I See It*, pp. 125f, 133, 135f, 173.
8. See especially, Karl R. Popper, *Conjectures and Refutations. The Growth of Knowledge*, London, 1963.
9. This means, as Polanyi points out, that scientific activity of this kind operates with an idea of knowledge based completely on identifiable grounds, *The Tacit Dimension*, London, 1967, p. 61. Cf. also *Knowing and Being*, London, 1969, p. 203.
10. Cf. F. S. C. Northrop, "Einstein's Conception of Science", in P. A. Schilpp, *op. cit.*, pp. 393–402.
11. A. Einstein, in P. A. Schilpp, *of. cit.*, pp. 21ff.
12. A. Einstein, *Out of My Later Years*, p. 61.
13. A. Einstein, *The World as I See It*, pp. 125, 136, 138; "Autobiographical Notes", in P. A. Schilpp, *op. cit.*, p. 13, etc.
14. A. Einstein, *The Evolution of Physics*, p. 30f; *The World as I See It*, pp. 134ff; *Out of My Later Years*, pp. 59ff; cf. Schilpp, *op. cit.*, pp. 131f, 140f, 247, 273, 279, 360ff, 372ff, 388f, 396ff.
15. Cf. Einstein's essays "Physics and Reality", *Out of My Later Years*, pp. 58ff; and "The Problem of Space, Ether, and the Field in Physics", *The World as I See It*, pp. 173ff.
16. M. Polanyi, *Science, Faith and Society*, new edition, Chicago, 1964, p. 87.
17. M. Polanyi, "The Unaccountable Element in Science", *Transactions in the Bose Research Institute*, vol. 24, no. 4, 1961, pp. 175–184; "Genius in Science", *De La Méthode. Méthodologies Particulières et Méthodologie en Général*, by S. Dockx et al., Brussels, 1972, pp. 11–25. See also the chapters on "Knowing and Being", "The Logic

THE SCIENCE OF GOD 95

of Tacit Inference", and "Tacit Knowing", in *Knowing and Being*, pp. 123-180.
18. A. Einstein, *The World As I See It*, pp. 128, 138; P. A. Schilpp, *op. cit.*, p. 81.
19. M. Polanyi, *Knowing and Being*, pp. 112-119, 129-137, 139-146. Einstein gave prominence to the sense of "wonder" in the origination of his thought. "For me it is not dubious that our thinking goes on for the most part without use of signs (words) and beyond that to a considerable degree unconsciously (*unbewusst*). How, otherwise, should it happen that sometimes we 'wonder' quite spontaneously about some experience? This 'wondering' seems to occur when an experience comes into conflict with a world of concepts which is already fixed in us. Whenever such a conflict is experienced hard and intensively it reacts back upon our thought in a decisive way. The development of this thought world is in a certain sense a continuous flight from 'wonder'". P. A. Schilpp, *op. cit.*, p. 9. Einstein then went on to give instances of "wonder" in his own experience, even at 4 or 5 and later at 12 years of age.
20. James Clerk Maxwell, *Scientific Papers*, edit. by W. D. Niven, Cambridge, 1980, vol. 2, p. 325.
21. A. Einstein, *The World As I See It*, pp. 148ff, 154ff, 156ff; *The Evolution of Physics*, pp. 125ff; P. A. Schilpp, *op. cit.*, pp. 25f, 33f.
22. A. Einstein,*The World As I See It*, p. 159.
23. Origen, *De Principiis, Origines, Grieschische christliche Schriftsteller*, edit. by P. Koetschau, Berlin, 1899/55, vol. V, pp. 32, 73, 132, 144, 208, 227f, 241, 273, 310, 322, 328, 345; cf. *Contra Celsum*, vol. VII, p. 193.
24. Clement of Alexandria, *Stromateis*, especially book VIII where he expounds the method of heuristic inquiry (*hē methodos tēs heurēseōs*), e. g., IV.16. For detailed references see my essay "The Implications of *Oikonomia* for Knowledge and Speech of God in Early Christian Theology", in *Oikonomia. Heilsgeschichte als Thema der Theologie*, edit. by F. Christ, Hamburg-Bergstedt, 1967, pp. 223ff.
25. Cf. here Michael Polanyi's concept of "indwelling" which is evidently developed from Johannine teaching in the New Testament. See *Personal Knowledge*, London, 1958, pp. 195-202, 279ff; *Knowing and Being*, pp. 134ff, 148ff, 160ff, 214, 220f, etc.
26. To think rigorously in accordance with the nature (*kata physin*) of God is necessarily to think in a "devout" (*eusebēs*) or "godly" (*theosebēs*) way.
27. See the useful edition by R. W. Thomson, Athanasius: *Contra Gentes and De Incarnatione*, Oxford Early Christian Texts, Oxford, 1971.
28. In both books, as the opening chapter of each makes clear, Athanasius was concerned to exhibit the inherent reasonableness of

the Christian knowledge of God in contrast to the "contradictions", "absurdities" and "irrationalities" of pagan thought. While he used logical reasoning to clear the ground of irrational ideas, it was another form of reasoning he used to establish the rationality of faith, through following the road taken by the truth in its own self-evidencing manifestations. *Contra Gentes*, 1f, 29.

29. *Contra Gentes*, 1.
30. Athanasius did not trace the rational order in the universe, which he described preferably in musical terms such as "symmetry", "concord", "harmony", "symphony", to an immanent rationality in nature but to the creative activity of the *Autologos*, the personal Self-Logos of God — *Contra Gentes*, 8, 34ff, 38ff, 42, 44. Along with the Hellenic separation between the sensible and intelligible realms he rejected the prevailing notion of a cosmological "logos" or the Stoic notion of seminal reasons (*logoi spermatikoi*) — *Contra Gentes*, 40.
31. *De Incarnatione*, 1–4. Cf. "The Hermeneutics of St. Athanasius", in *Ekklesiastikos Pharos*, Addis Ababa, 1970, vol. 1, pp. 89ff.
32. *S. Anselmi Opera Omnia*, edit. by F. S. Schmitt, Edinburgh, 1946, vol. II, pp. 37–133. See *Cur deus homo*, 1.1, vol. II, pp. 47ff.
33. *Cur deus homo*, 2.19, vol. II, p. 131. See also *Epistola de incarnatione*, 1, vol. II, p. 7, where he challenges an abstraction of the understanding or the structure of the reason from the object of knowledge and then a deployment of it in speculative argumentation outside the field of actual experience.
34. *Cur deus homo*, 2.9, vol. II, p. 106; 2.19, vol. II, p. 131; *Monologion, prol.*, vol. I, p. 7f; 19, vol. I, p. 34. For detailed references see my paper "The Place of Word and Truth in Theological Inquiry according to St. Anselm", in *Studia mediaevalia et mariologica P. Carolo Balić septuagesimum explenti annum dicata*, Rome, 1971, pp. 133–160.
35. *Cur deus homo, Praefatio*, vol. II, p. 42. St. Anselm claimed that only when we can embody in our thinking a rational order or sequence which reflects or is controlled by the *ratio veritatis*, or the inner logic of the subject-matter, do we engage properly in scientific activity. *Cur deus homo*, 2.19, vol. II, p. 131; *De concordia*, 1.2, vol. II, p. 248.
36. Since theological knowledge is grounded upon the same truth upon which the Holy Scriptures are grounded, and which is what it is independent of them, theological understanding begins where the biblical statements cease. This appears to reflect the same point cogently argued by Hilary of Poitiers in the *De Trinitate*, II, 2, 5.
37. Anselm works this out in the *De veritate* rather carefully, vol. I, pp. 176–199. Cf. my analysis of this in *Theologische Zeitschrift*, vol. 24, 1968, pp. 309–319.
38. Anselm's claim to operate *sola ratione* is, therefore, far from being

rationalist, for we exercise our reason only in fulfilling its obligation or indebtedness (*debitum*) to the Truth of God the ultimate *ratio veritatis*, that is, through humble submission or obedience to it.

39. *Philosophical Fragments or a Fragment of Philosophy* by Johannes Climacus, tr. by David F. Swenson, Princeton, 1936.
40. Cf. in this respect Hermann Diem, *Spion im Dienste Gottes*, Frankfurt, 1957. It is significant that whenever Kierkegaard speaks of "paradox" or the "absurd", he never does so under his own name but always under a pseudonym which he has adopted in order to outflank and break up a misunderstanding that has arisen through a false or refracted relation to the truth. See A. McKinnon, *International Phiosophical Quarterly*, vol. ix, no. 2, June, 1969, pp. 165–176.
41. A. Trendelenburg, *Logische Untersuchungen*, Berlin, 1840. See *Concluding Unscientific Postscript*, tr. by D. F. Swenson and W. Lowrie, Princeton, 1941, p. 100. Cf. H. Diem, *Kierkegaard's Dialectic of Existence*, Edinburgh, 1959, pp. 18f, 32f, and Paul Sponheim, *Kierkegaard on Christ and Coherence*, London, 1968, p. 67 who cites Kierkegaard as saying: "There is no modern philosopher from whom I have had so much profit as Trendelenburg."
42. S. Kierkegaard, *Philosophical Fragments*, pp. 44ff, 60ff. Cf. *Concluding Unscientific Postscript*, p. 306.
43. *Philosophical Fragments*, pp. 60ff; *Concluding Unscientific Postscript*, pp. 277ff.
44. *Concluding Unscientific Postscript*, pp. 91ff, 96, 231, 306, 327, 340, 343.
45. *Philosophical Fragments*, pp. 47ff, 67ff, 83f.
46. James Clerk Maxwell, *A Treatise on Electricity and Magnetism*, Dover edit., New York, 1954, vol. 1, pp. 9f; vol. 2, pp. 176f, 196ff, 209f, 211ff, 247.
47. S. Kierkegaard, *The Concept of Dread*, tr. by W. Lowrie, Princeton, 1944, p. 12. Cf. the long discussion in *Concluding Unscientific Postscript*, pp. 99ff, 273ff, 279f, 296, 313. And see H. Diem, *Kierkegaard's Dialectic of Existence*, pp. 18f, 27f.
48. Cf. A. Einstein, *The World As I See It*, p. 159; and "Autobiographical Notes", P. A. Schilpp, *op. cit.*, pp. 27, 37.

CHAPTER 4

THE SOCIAL COEFFICIENT OF KNOWLEDGE

IN the last two chapters we examined the rational form which knowledge of God takes in our human understanding, and I tried to elucidate the connections between form and objective content by comparisons taken from the relations of geometry and logic to knowledge in other fields of scientific investigation. Since theology can be pursued only within the space-time structures of this world, the conceptual form of theological knowledge cannot be sharply segregated from mathematical or logical form, so that this interconnection between theological science and other science is not by any means irrelevant. We may begin this chapter, however, by employing another comparison taken from the realms of *art*.

We now move to a different kind of form, still symbolic form, but *suggestive form*, yet form which like the geometrical and logical forms which we have discussed, is partly the product of our creative imagination and partly the product of objective features and properties of reality. Here we have to do with a profound aspect of reality in its ultimate harmony, unity and simplicity, which outruns the explicit and formalised forms which arise in us under its pressure upon our consciousness. Our universal mode of expressing form in this sense is *art*, whether in painting, music or literature, but it is a basic element in all rational form, as we can discern in the beauty of a scientific theory, or the elegance of some argument, or in the beauty of holiness. The form I am concerned with now is form at its most sensitive point where its "antennae", so to speak, are in touch with ultimate reality, form in its transcendental reference to the beyond, to the unformalisable *forma formans* which acts creatively upon us, not to reproduce itself in our formalising activities, but to call them, as it were, into contrapuntal sequences and patterns of an open texture through which it can reverberate or resound in the human spirit. Hence our attention will not be directed to art-forms as such but to the kind of *semantic intentionality* that characterises them, and with the orientation in the human consciousness as it is pointed in this way to the intelligible ground of reality.

We have something here not unlike what the physicist is concerned with in the unspecifiable correlation between his symbolic notations and the comprehensibility of the universe, or in what a biologist like Pantin is grappling with in what he calls "the illative sense", borrowing a telling expression from J. H. Newman,[1] or an astrophysicist like C. F. von Weizsäcker when he speaks of "meditation" as the means he employs to raise his thought transformally to a higher level,[2] or again what Polanyi describes as "heuristic vision", taking his cue from Christian worship,[3] but it is this semantic intentionality that gives meaning to the whole framework of human life so that without it every culture slips away into meaninglessness. It is for this reason, I believe, that as we progress in our manifold scientific inquiries into the harmonies and symmetries of the universe, and as the inner relation between what we know and our knowing of it deepens, there presses hard upon us for realisation within the personal and social patterns of our human life measures of correlation with the regularities and invariances that are disclosed in the frame of created reality. And it is in this area that theological knowledge, having to do with the relation of God to the world, traces a covenanted relation between the creation and the constancy or faithfulness of the Creator. And so we come to what I have called the social coefficient in human patterns of behaviour to the knowledge of God, patterns which *are or should be* correlated with the invariances of the creation as it unfolds its rich and manifold order to our inquiry.

I have said "are or should be", as we have serious

problems in this regard which come to the forefront especially in transitional periods when the social structure of human life struggles with adjustment to the insistent demands of intelligible reality, and is not infrequently found in disengagement and flight from the self-criticism and discipline that knowledge of reality brings. This is what I think we can discern everywhere in the modern world, in the widespread reaction from pure science, for example, in obsession with merely formalistic or technological patterns of thought, or in emotive and existentialist and relativist views of ethics. Before I proceed to discuss the social coefficient of knowledge, let me indicate what I mean by this reactionary disengagement and flight from the constraints of objective reality, by drawing out a little the analogy from art which I mentioned above.

In the romantic period of modern art, in literature, painting and music (with natural differences of course in each sphere), there took place a considerable shift in orientation away from the intentionalities of our existence toward the ultimate ground of reality, to the view that the content of art is predominantly subjective and personal. Then there took place a move in another direction to a view of art in which explicit expression of positive content is replaced by abstract symbolic systems built up of signs which supposit for, rather than intend, reality, external or internal, and which are manipulated according to certain accepted rules of transformation. There is an obvious connection here between these movements in art with the romantic and positivist attitudes to the world which we have been passing through. In both cases, however, the referential and intentional relations with reality were cut or damaged, or at best were only of an oblique sort, so that meaning came to be located (a) in the personal or subjective coefficient governed by self-expression and self-fulfilment, or (b) in a play of imaginative symbolisms which could derive significance only in the actual context of the society out of which the artist projected his cyphers. Both of these tendencies were certainly reactions against the kind of art that is primarily concerned with a mimetic congruence with reality. But both seem to reflect a world in which man has lost his bearings in the universe and both seem to be basically empty of meaning for they cut themselves off from the significant constraints of the real world and from the emergence of meaning as we attend away from ourselves to the universe around us. They try to create meaning for themselves in ways in which meaning is bound to elude them.

Manifestly art by its very nature is not concerned with explicit formalisation or a one to one correspondence with reality, for it depends on an element of artificiality or even of estrangement from nature.[4] To be a work of art something must be able to set the human imagination free from its imprisonment in the mimetic forms in which we are inevitably implicated, and thus provide the occasion for those appreciating it to transcend the limitations of their own place in space or time. But when the imagination becomes completely detached from the compelling claims of our actual existence in space and time, it becomes merely a meaningless dream, indulged in for its own sake like a fanciful game. A genuine work of art must have a grip upon reality in its depth, while declining to reduce that grip to explicit formalisation, and so by its nature indicates far more than it can imaginatively depict to people at the time. In this way, however, a work of art is so full of meaning that it commands a universal range of appreciation in time as well as space, so that though it may have been produced in some ancient culture it continues to be appreciated and understood, for it lifts the mind of each generation to a level of reality that is invariant for each and every generation.

I am not concerned at the moment, however, with art as such, but with the problems in the relation of art-form to reality, for it is in this extremely sensitive area that we have to do with these informal, creative and spontaneous movements of thought which are so basic and all-important in every area of knowledge. Nor am I concerned with the second of the problematic tendencies which I have mentioned, that reflected in abstract art, for I have already discussed sufficiently in earlier chapters the positivist and abstractive science to which it is akin. What I am concerned with is the problem, forced on us by the modern world with such renewed urgency, of the correlation between our personal and social structures and the intelligible grounds of reality, that is, with the cultural milieu within which our understanding is nourished, and within which new thoughts and fresh glimpses of reality are born. Is this frame of personal and social existence *semantic as a whole*? Does it have meaning as an entire framework? If it does not, how can it nourish meaning for us in this or that area of knowledge, and how can it be the matrix for those essential clues which are the *sine qua non* of all scientific and rational inquiry? There is an inescapable need for a social coefficient of knowledge in order to establish and maintain semantic relations with reality, within which man can be at home in the universe, through being rightly related to its essential patterns and intrinsic intelligibilities which are the ground of meaning. It is within this framework that human conceptions are constantly formed, patterns of thought take shape, and the anticipatory grip upon reality which initiates inquiry is gained. But of course our special concern is with the *social coefficient of theological knowledge*, and with the way in which our basic theological concepts arise in the dynamic and empirical correlation of our human life to the self-revealing interaction of God with us in the world.

A social coefficient of knowledge can be analysed only *a posteriori*, that is by looking back after the attainment of knowledge. It arises in the general symbiosis between a society and the world of intelligible reality, or between a particular community, scientific or religious, and its chosen field of interest, but it can have no meaning or validity when it is isolated from what is known, and treated in itself, for it is not something that exists in its own right or that can contribute knowledge through its own functioning *a priori*.[5] It is to be granted that nothing can be apprehended apart from the synthesising and conceptualising activity of the human reason, which seems to have been the root idea behind Kant's "synthetic judgments *a priori*". However, as we found in the first chapter, it is quite another thing with Kant to define what is possible cognitive experience in terms of our ability to construct it, as though any proper object of human knowledge were a construction which we made out of space and time, which sooner or later would become a construction out of our consciousness. No more can the social coefficient of knowledge, regarded as a corporate counterpart to the synthetic *a priori*, be identified with some synthetic or constructive force in the social consciousness, and as such be open to analysis apart from actual experience and knowledge. There can be no social coefficient of knowledge in that sense, any more than there can be final categories in the sense of Kant.[6]

If we may refer to the subjective coefficient of knowledge as the capacity of the subject to be affected and modified by what it knows independent of itself, we may refer to the social coefficient of knowledge as the capacity of a society or a community to be affected and modified through its advance in knowledge of what is independently real. It must be noted, however, that here we have to do with an *affect*, not an *effect* from which we could argue to a specific cause and thereby derive explicit knowledge. This implies that the social coefficient of knowledge is an affected modification of our social consciousness from which we cannot derive specifiable epistemic content.

On the other hand, we must not overlook the fact that the power of the subject is also its power to grasp objective being, to receive the effect of its intrinsic intelligibility, recognise its intimations and interpret the clues to which it gives rise. And that in turn indicates that in its capacity to be affected and modified in the orientation of its social consciousness toward objective being, a society or community provides the semantic frame within which meaning emerges and is sustained, but as such it functions through suggestive reference in the mode of art-form rather than in the mode of explicit mathematical or logical form. Regarded in this way the social coefficient of knowledge constitutes from generation to generation the sensitive matrix within which our all-important informal relations with reality are evoked, and thus constitutes the medium in which those relations while not formally communicable are nevertheless communicated through common participation in the affected modification of social consciousness. Just as each of us comes to know far more about the physical or moral order of the world than we can ever tell by the time we are five years of age, so we early acquire an ability to read what is engendered in us by the structure of social consciousness in which we grow up, to grasp signification and intention, and are thereby disposed to recognise the significant manifestations of reality and open our minds to its disclosures. It is as we are nursed and trained by the social coefficient of knowledge embodied in the society or community to which we belong, that we also gain the power of judgment to relate experience to patterns of meaning, and then the initial acts of recognition develop into acts of identification which complete the process of inquiry in which we come to engage.

Now, of course, this is always the point of possible distortion or error. We normally identify something by assimilating our interpretation of it to some pattern of coherent meaning in virtue of which we can make positive and coherent statements about it. But when we bring into play some prior structure of meaning or system of ideas which has been conditioned by concepts derived from elsewhere with little relevance to the reality in question, distortion easily takes place. Distortion of this kind arises quickly whenever the social coefficient of knowledge is turned in on itself and is treated as man's social self-understanding which he uses as a regular guide in his interpretation of facts or events in any realm of inquiry.

Hence, even when we turn to analyse *a posteriori* the social coefficient of knowledge we must be on our guard against letting it fall under the control of some prior, or even *a priori*, self-understanding which people may claim for themselves apart from any modification by empirical knowledge — that is, in our case, apart from actual experience and knowledge of the living God. Difficulties are bound to arise from the fact that the knowledge of God with which we are concerned is knowledge by persons, and persons in community, who can go on existing as such, engaging in highly rational and personal activities, apart from their knowledge of God, at least in any explicit sense. This reminds us that in analysing the social coefficient of theological knowledge we must take into account the fact that it cannot be cut off sharply from the social coefficient of a range of knowledge far beyond the bounds of the people of God or the Church where the distinctive coefficient of knowledge of God is embodied. From within the purview of theological knowledge itself, however, we have to reckon with the fact that God reveals himself to us in such a personal way that he establishes relations of reciprocity between us and himself and calls forth a community of reciprocity — the people of God or the Church — which he assumes into covenanted partnership with himself. In making himself known to his people he creatively evokes their corporate responses and harnesses them in the service of his continuing self-communication to mankind. That is to say, God includes a personal and social coefficient on our part within the structure and operation of revealed knowledge of himself. But in the nature of the case this coefficient cannot be abstracted out of that relation of reciprocity in order to be analysed as though it had an autonomous structure or an independent force of its own, far less to be treated as some sort of epistemological *a priori*. Nor, of course, may it be interpreted in the light of any kind of pre-understanding which we may generate independently of the actual field in which knowledge of God arises. Nevertheless, we must take into constant account the fact that all knowledge on man's part is socially, culturally and historically conditioned, so that we have to reckon constantly on a tension between an adequate social coefficient that arises out of actual knowledge of God mediated through his self-revelation and a social coefficient that has as a matter of fact arisen on other ground in other areas of human experience.

How, then, may we speak of this personal and social coefficient in the light of the actual knowledge of God with which it arises? In making himself known to us God discloses at once both that he has made himself open to us and accessible to our knowing of him and that even in revealing himself to us he remains ineffably transcendent over our knowing of him.[7] The counterpart to this, with which we are concerned in the personal or social coefficient of knowledge of God, may be regarded as an openness or a readiness in man for God, but as an openness to the infinite and eternal reality of God it has a distinctly indefinite, inarticulate character.[8] In the reciprocal relations God establishes with us it is the creaturely, contingent *vis à vis* to the unlimited reality of God who is greater than we can ever conceive,[9] the disposition given to our minds in shared response to God's self-revelation as through communion with him we are elevated, as it were, above ourselves.

It must be stressed that this is not an independent openness or readiness traceable to some immanent property or autonomous structure in human and social being, which can be examined and exploited as the ground for an independent conceptual system of "natural theology" in the sense in which we had to dismiss it. It is bound up, however, with the contingent nature of man and his intelligibility which in their utter difference from God are entirely dependent on him. But that is a truth of man's creaturehood which we may understand and regain only in the light of God's self-revelation as our Creator. Within the bi-polar relation of theology, then, which we discussed in earlier chapters, we may think of this openness or readiness for God that characterises the social coefficient of knowledge of God as an affected modification within the creature-Creator relationship. It may be understood as the inner determination of the creature through the freedom of God to be present to him, to uphold him in his contingent relation to the Creator and to bring that relation to its fulfilment in him, and thus to make the creature open for knowledge of God beyond any capacity he has in himself.[10] Expressed somewhat differently, the social coefficient relates to the communion which God grants his people through the presence of his Spirit, and may be spoken of as the adaptation toward divine self-disclosure that arises in that communion or the capacity for contuitive recognition mediated in the actualisation of revealed knowledge of God. Is this not what we mean when we speak of man's *spirit* in relation to *God's Spirit*, and thus of his spiritual discernment within the fellowship of the Church?

At an earlier point we noted that when we seek to analyse the social coefficient of theological knowledge *a posteriori* and decline to treat it as an independent structure unaffected by God's actual revelation of himself to mankind, nevertheless we must not forget that it arises within the general frame of human society in its continuing interaction with the intelligible world in which we live, and cannot escape being conditioned by it. Hence we have to reckon with a critical relationship between the distinctive coefficient of our knowledge of God and social coefficients of knowledge that develop in other areas of human existence and experience. That is the angle from which we must now consider our subject.

In his Preface to *The Critique of Political Economy* Karl Marx wrote this: "The mode of production of material life conditions the general process of social, political and intellectual life. It is not the consciousness of men that determines their existence, but their social existence that determines their consciousness."[11] There we have expressed a very fundamental point in the Marxist concept of social reality: it is man's social existence that determines his consciousness, which implies that all our basic beliefs and convictions have an essentially communal character. According to Marx, however, even man's social existence is so deeply interwoven with material activity, that human conceiving and thinking are the direct efflux of material behaviour, and are conditioned by the processes of material production. Thus the whole system of human thought, including the development of science, is to be regarded finally as the superstructure of underlying forces of production.[12] Here we have wedded together the concepts of human society and technological rationality. There is no such thing, therefore, on this basis as pure science concerned with reality for its own sake, resting upon given objective truth, rooted in and correlated with a free society.

Now in contrast let us consider the Christian Church. In its classical form it is not regarded as a man-made but as a divine institution. It is the community that arises within our social and historical existence as the social correlate of an intelligible realm which we do not create but discover and to which we are committed in responsible processes of life and faith and thought. Here we have wedded together the concepts of human society and an intrinsic rationality of an infinite depth and range of objectivity which has given rise to the open structures of life and thought grounded transcendentally beyond ourselves. This yields the concept of the person as the bearer of objective value,[13] and the concept of the free society of persons within which science and theology are pursued unfettered by closed ideologies.[14]

Both Marxism and Christianity share the view that our knowledge is rooted in the social structure of human being, but they are poles apart in the way in which they undertand the communal character of our basic beliefs and convictions to arise and develop. For Marxists the coherent structure of man's intellectual life is traced back to economic determinism at work in the material processes of life and history, which throws up a positivist conception of science demanding a positivist programme for the ordering of society. "The true fulfilment of such a programme", as Polanyi has pointed out, "is the destruction of the free society and the establishment of totalitarianism."[15] For Christianity, on the other hand, the coherent structure of man's intellectual life is interpreted as an organisation in human being arising under the impact of a spiritual reality to which we are summoned to respond through spontaneous reconstructions of our historical existence, and which we progressively understand as we commit ourselves to the claims of its inherent intelligibility and develop our science through free and open inquiry in the light of it. The theoretic structure of our science and the rational structure of society are coordinated with one another, through common dedication to the service of a transcendent reality. This is very evident in a university, for example, where the structure of an academic community and the structure of scientific inquiry develop in correlation with one another in their service of truth.

It is our purpose here to consider the social coefficient of knowledge in this coordination between the communal character of our convictions and the harmonious structure of the intelligible universe, for it is within that coordination that we deepen our understanding of the way in which concepts are coordinated with experience, in the manner in which they arise and in the manner in which they are to be verified.

In all authentic knowing we distinguish what we know from our knowing of it and at the same time we distinguish ourselves from whatever we know. We recognise our own free independent existence and we are aware of ourselves as rational subjects in the activity of knowing. But obversely we recognise what we know as having reality "on its own", independent of our knowing of it. In distinguishing ourselves from what we know we are aware of ourselves as irreducibly real subjects, who have reality in ourselves independent of other realities with which we stand in relation. But by the very same token we are aware of the other as having reality in itself independent of our knowing of it. It is this personal mode of being as subject which is precisely the mode of being in which we are aware of the objective world around us. Personal subject-mode of being is thus the bearer of objectivity.[16] In being aware of ourselves as subject-beings we are aware that we act from a centre in ourselves, with our own source of activity as rational agents, not controlled from beyond ourselves, and we resist any attempt from beyond ourselves to manipulate us. Obversely we are aware of others as resistant to us, who act from a centre in themselves as rational agents or subjects. That is to say, they are objective to us and they object to any control which we may seek to exercise over them. They have an inviolable character, something sacred to themselves alone, which we must respect. In a certain sense, this applies also to object-beings, and even to our own artifacts, such as a motor car, for we may not use it for flying in the air or engaging in submarine exploration. We have to respect it for what it actually is and act in accordance with what it actually is, that is, respect it objectively for its own sake.

It is only through this kind of interaction between ourselves as knowers and what we know that we know. But instead of obstructing knowledge of things as they are in themselves, this kind of interaction must be construed as the steps we take in maintaining contact with what we know. Hence the subjective variable that enters into our contact with reality must be interpreted as a necessary step on our way toward objective knowledge of it. It is a transaction with being. The personal relation, then, that enters into our knowledge of being is one in which we attend to it in and for itself and not one in which we value it for our own subjective satisfaction. This is precisely the mode of personal relation which we cultivate in community, in inter-subjective social existence, in which persons are rational agents in interaction with other rational agents, each interacting objectively with the independent free reality of the other.

Thus personal being is, I submit, the prime bearer of objectivity, for that kind of relation is the relation in which persons as persons are. What we mean by personal being is precisely that kind of being which by its nature is oriented beyond itself, in the other, in God ultimately, and in other human beings relatively, that is, in other personal beings. Hence the person cannot be defined through exclusive reference to itself but through its relation to other persons, i.e. objectively. Our human existence as embodied subjects or agents, in which we use our bodies to make contact with other bodies, distinguishing them from ourselves, enables us to transcend the limits of our individual existence, and to participate in a realm of interconnection and communication with other rational agents which far outreaches the bounds of our own limited experience and understanding. But for that very reason our understanding is always being stretched beyond what we have already apprehended, so that we learn to operate, as it were, by reaching beyond our original grasp. In this way we are aware of ourselves as belonging to a realm of spiritual reality in which we are in the way of knowing more than we can say, while even our statements made in this inter-personal or social milieu indicate more than we can ever specify at the time and are far from exhausting the realities they intend. The spiritual reality to which we belong has a range of content which we cannot infer from what we already know, but which we may get to know more fully only through heuristic acts of exploring entirely new ground and grappling with novel connections and ideas.[17]

Our inter-personal relations with one another have thus an open-ended, transcendent relation built into them, which is constitutive of our personal and social reality. It is that same open-ended or transcendent orientation which we bring to our relation with whatever we seek to know. We know only as we break through the isolation or enclosure of ourselves and allow external reality to disclose itself to us; while we on our part openly assent to what it is under the compulsion of its being what it is and not something else. Hence intensely personal acts of relation, discernment and judgment belong to the epistemic act in every field of rational knowledge and fundamental science.

On the other hand, we must not allow ourselves to forget that, although knowledge involves a relation between the person of the rational agent and what he knows, in authentic knowledge we distinguish ourselves and our knowing from what we know. Hence far from projecting our subjective states and conditions into the content of what we know we build into our sciences devices to prevent that from happening. This is perhaps why the word "I" does not find its way into books on science (in interesting contrast to a mode of writing of some modern philosophers!), for the scientist has no romantic interest in expressing his own individuality or asserting the place of his own personality in knowledge.[18] But this is not meant to argue that science may be pursued in a ruthlessly impersonal, disinterested, detached way, that is, objectivistically, as if the knower himself had no part in the activity of knowing, for, as we have seen, it is precisely through personal acts that we make contact with intelligible reality and are able to recognise and interpret the informational content of its disclosures.

We come back to the point, then, that it is the interpersonal structure of our social existence, our openness as persons to one another, in which we share experience of reality and reach communal convictions about it, that develops in us the modes of cognition which enable us to engage in objective inquiry and reach objective results. The really objective is that which is shareable, what we can experience together or in common, and which is transcendent to each of us and therefore also to all of us. Hence in all objective knowledge we try to eliminate those features which we cannot share together or cannot have in common with others, and which are relative only to this or that person in this or that particular situation.[19] It is the closed mind, the autistic individual, the detached person, who is incapable of communicating with others, who is obstructed in the pursuit of knowledge or in the prosecution of scientific inquiry. This also applies to a society or community that has become inbred in its own ideological development and uses its own objectified self-consciousness as the criterion for reality or authenticity.

It is to this social sub-structure of knowledge that we may relate what Michael Polanyi has called the inarticulate grasp of reality or the tacit dimension upon which all rational knowledge and all scientific inquiry rely.[20] It arises out of an intimate experience of the real world which we share with one another. In itself it is a non-formal apprehension of reality, but it constitutes the necessary ground or condition for all explicit knowledge such as we develop in the various sciences. It is possessed of significant and suggestive properties in virtue of which it not only underpins and guides constructive formalisation of knowledge but projects our thought heuristically beyond the empirical evidence of our knowledge in the present, for it is geared into an unbounded intelligibility which invites unceasing inquiry and unconditional commitment. It is surely in this light that we are to appreciate the community structure of our human life and the moral ordering of our affairs as they press in upon us for ever deepening realisation in obedience to the imperatives of ultimate reality. This state of affairs, however, raises the question of a transcendent source of rationality and the activity of an ordering Agent which we considered in the second chapter and to which we shall return in a later chapter.

Here, then, we proceed on the assumption of a deeply significant correlation between the structures of our social consciousness and the harmonious structures of the universal nature of things which is inexhaustibly objective and transcendent to our grasp of it whether in inarticulate or articulate frames of understanding. This is not to claim that there is an inevitable and necessary connection between the structures of our social consciousness, or the frame of our communal convictions, and the inherent lawfulness of created being, but only that they are *affected* by it and are therefore liable to refraction or distortion. But that is a state of affairs which imposes on us the obligation constantly to clarify those structures and convictions in tracing them back to their creative ground. Certainly the formalisation in an explicit rational framework of our inarticulate grasp of reality enormously deepens and strengthens our grasp of it, but we must always remember that no cognitive structures on our part, social or scientific, must yield knowledge of reality. Thus the social coefficient of our knowledge is to be interpreted in terms of that astonishing coordination between our personal and social consciousness and the inherent intelligibility of the universe which provides the semantic focus within which there arises in us the ability to recognise and identify its various patterns, for it is only through incipient intimations or clues that develop in this way that we may initiate and carry through any proper scientific inquiry and advance the explicit content of human knowledge.

This social coefficient of human knowledge has immense advantages but it also has a serious drawback. On the one hand, it enables us to be *at home* in the universe, not strangers but those who belong there and who are in some intuitive way familiar with its reality. It is in this respect that the web of meaning that is found lodged in human language is so significant, for language, as Heidegger used to say, is the house of being, and it is through language that reality shows itself to us and we become familiar with it.[21] This function of language is basic to what we have been calling the social coefficient of knowledge, as it enables us to be rightly related to the intelligible universe around us so that it can shine through into our understanding and we are not left to grope our way singly in the dark. Thus in our inter-personal and social existence we find ourselves caught up in a comprehensive outlook without which we would be rather blind and would not be able to recognise or identify significant aspects of reality. Apart from the proleptic glimpses of reality in its inherent patterns which arise in our minds in this way, all our inquiries would be undirected and uninformed and therefore could do little more than make useless shots in the dark. It is worth repeating at this point that the social coefficient of our knowledge, or the cognitional structure of our social consciousness, does not generate in us concepts of reality, nor does it provide our knowledge with informational content, but it does predispose us toward explicit apprehension of the rational order intrinsic to the nature of things through the informal, inarticulate way in which it reflects it.[22]

On the other hand, the social sub-structure of our thought can become self-contained. It can grow in upon itself and thus lose its openness to the vast intelligibility of the universe. It has the tendency to develop an independent life and momentum of its own, and to assume power in prestructuring our life and thought. Instead of opening a door upon reality for us and making us ready to recognise its significant disclosures it can assume a hard paradigmatic, prescriptive role which blocks the way of advance in knowledge. This can happen when under pressure from below the knowing relation becomes inverted and the creative source of intelligibility is located in the human consciousness itself instead of in objective reality when it takes on a categorical and absolute character which cannot be modified by further experience. A fixation or rigidity of this kind sooner or later leads to the stultification of rational or scientific inquiry. But, as we have already seen, this is what happens when the technological rationality gets out of control and a positivist conception of human thought imposes a deterministic clamp upon the spontaneous development of our intellectual and spiritual life. Apart altogether from the deliberately anti-metaphysical programme of certain philosophies, this tendency seems to be steadily at work in our social consciousness, manifesting a sort of creeping paralysis, so that we need to find ways of counteracting it, if we are to break loose from its grip and regain the natural freedom which our minds have before the vast incomprehensible comprehensibility of the universe. It is a sad fact that the artificial environment in which we so often live affects our powers of perception in such a way as to make us incapable of seeing things as they really are, so that if we really are to advance beyond what we already claim to know we must be freed from our own fetters and transcend the bounds of our current operational framework of thought. But since the articulate framework of thought with which we operate is grounded in the ontological substructure of our social existence, it is that social existence that needs to be changed if any new outlook is to be achieved, or we are to advance to radically new knowledge. Thus while this structure of our inter-personal and social consciousness constitutes an all-important coefficient in knowledge, it is of no use to us unless it functions in such a way that we are always breaking through it and transcending it into objective realms beyond. The cognitional structure is not and cannot itself be a source of our concepts but only of openness and readiness for what is new, if it is properly related to reality.

How then does this happen? How do we break free from the social structures which regulate and shape our intellectual development from below, and so gain new insights into a larger whole of which the old framework of thought is found to be only a partial and distorting reflection? How do we enlarge our perspective so that we may grasp and assimilate radically novel relations and novel forms of thought? This is certainly not possible merely by shattering the old frameworks of thought, as so many people today seem to advocate, the nihilists and anarchists, the "Marcusans" and the so-called "new left", in the rather irrational hope that something "better" will quite arbitrarily emerge, for we cannot even begin to advance toward the new without some guidance derived from the framework of the old. And yet we have to agree with them that it is only in a struggle with the social structures in which we are actually implicated that we can become free for the new.

In several of his books Michael Polanyi has argued, against positivist and Marxist notions of science and the kind of technological society they imply, for dedication to the service of a genuinely transcendent or spiritual reality if pure science, unfettered by the dictates and dogmatisms of the technological society, and therefore scientific freedom, are to be preserved and encouraged.[23] By transcendent reality he referred not only to what is independent of our knowledge of it and which can be known only out of itself, but the kind of reality which presses for realisation in our minds while continually breaking through the limits of our conceptions and descriptions and which holds out to us the promise of an indeterminate range of future disclosures far beyond what we could expect or anticipate. It is in submission to the transcendent or spiritual reality of truth over which we have no control, in acknowledgment of transcendent obligations and in the dedication to transcendent ideals, which we cannot but affirm to have universal force and validity, that we have our freedom as rational beings, not only as individuals but as members of a community or society. We need, as it were, an Archimedean point far beyond us, and indeed beyond the world, through which we can be levered out of our rigid fixations and social mechanisms, and liberated for the pursuit of pure science concerned with reality for its own sake, and for the free and open-structured society that is correlated with it. Science, faith in transcendent reality, and the free society are inseparately interlocked together.[24]

It is not enough for us, however, to have in view the kind of transcendence we have to do with *in* the world, that is, with objective realities possessed of intrinsic intelligibility, which we cannot subdue to the mastery of our scientific conceptualising and formalising operations. We need to have in view the infinite transcendence of God *over* the world, who, as the Creator of the universe and its immanent rational order, is the ultimate ground of all transcendence within the universe and to whom the universe as a whole is left open at its boundary conditions. It is, of course, with the Transcendence of God that we are primarily concerned here. He is the one Archimedean point beyond the universe to whom the universe as a whole is so related that it is given authentic meaning throughout all its immanent structures, and the one Archimedean point to whom we are so related within the universe and all its science and social structure that we are constantly emancipated from ourselves and enabled to transcend the structures of our scientific and social activities.

This brings us to the Christian Church as the social correlate of God's self-revelation, the inter-personal community of on-going symbiosis between God and society. The Church is the community in the midst of our social existence committed to a transcendent reality, the ultimate ground and reason for its very existence. As the community freely called into being through the Word of God and sustained throughout its life and activity in space and time in covenanted partnership with himself, the Church must be regarded as correlated to the infinite depth of hidden yet partially accessible truth far exceeding the capacity of any one person to fathom and in respect of which the whole Church throughout the ages is a far better but still far from adequate correlate. This Church should then by definition be continually and unrestrictedly open to the infinity and eternity of God, and as such is meant to lift the horizon of mankind toward his transcendent Reality and provide the semantic focus within which faith and intuitive contact with God may spring up and yield ever-deepening understanding of him. It is through this unbounded openness to the infinity and eternity of God that the contingent structures of the Church in space and time should take shape and thus reflect the light of the transcendent Reality of God and the meaning of the finite but unbounded universe which he has created.

That is the kind of community in which the theologian is called to serve God's self-revelation to mankind and articulate within our creaturely forms of thought and speech the knowledge which God gives us of himself. It is within the context of worship and wonder in the Church and within the fellowship of intimate experience of God, that his mind is informed by intuitive apprehension of God, without which he would remain blind and undiscerning in the world, and unable properly to interpret the evidence before him. It is thus through the common tradition of shared spiritual experience and insight in the Church that the theologian makes cognitive contact with the hidden realities of God into which he inquires, and then through following up the clues thus mediated to him he seeks to evoke depths of knowledge not discerned hitherto. By developing appropriate intellectual instruments with which to lay bare the underlying epistemological pattern of thought, and by tracing the chains of connection throughout the coherent body of theological knowledge, he feels his way forward to a more precise understanding of the ways and works of God in the hope of deepening the Church's grasp of the divine Reality. Faith seeking understanding, always in the mode of prayer in response to the disclosures of God's Word, and the sustained discipline of hard mental activity, are interwoven in his inquiries, but they are not isolated from the fellowship of God's people and bear fruit only within the objective perspective opened out through the mind and tradition of the Church as it is unceasingly correlated to the creative intelligibility of God.

Let us think of the Church, then, as that which results from the intersection, so to speak, of the vertical and horizontal dimensions, that is, what the writers of the New Testament spoke of as *koinonia*. Koinonia refers to the common communion which we have with God through Christ and in his Spirit, but it also refers to the communion of love which we have with one another on that basis. In *koinonia* the community of reciprocity which we have with God is actualised within the reciprocities we have with one another in human society, but in such a creative way that our reciprocities with one another are organised and informed by the intelligible presence of God through his Word and Spirit indwelling the Church, and are at the same time deployed in the service of God's love and will for all mankind.

Now undoubtedly, the intersection between the vertical and horizontal dimensions, the actualisation of the divine society within the life and structure of human society, gives rise to an ambiguous and problematic state of affairs. The Church's consciousness of God is not unaffected by the social consciousness of people in which the Church is embodied. Its understanding of God is refracted through its involvement in the world's understanding of itself reflected in human culture. Divine revelation is grasped and interpreted through the inarticulate framework which underlies the structures and institutions of our day to day inter-personal existence as well as through the articulate frameworks of thought with which we operate at the time. It is understandable, therefore, that the corporate relation which we have toward God in and through the Church and the development of our understanding of his revelation in the tradition of the Church, can easily be distorted and diffracted under the conditioning of non-theological factors deriving from this or that passing phase of culture, with the result that there arise different interpretations of divine revelation and multiple patterns of Church order, conflicting with each other. This is one of the roots of the divisions in the Church, as instead of overcoming or transcending the many divisions that have arisen in the life of mankind, the Church has allowed them to eat back into itself and call in question its foundations in the reconciling and unifying Love of God. These problems are particularly acute today owing to the pluralistic fragmentation of society and the disintegration of form in the arts which have overtaken modern life.

This is not, of course, a problem peculiar to the Church, but it is a particularly difficult problem for the Church, just because it is charged with the message of the reconciling and unifying Love of God in Jesus Christ, and because it must communicate that message to people in the language and culture of their times, that is, within the framework of their social consciousness, so that it is constantly tempted to adapt its interpretation of the message to the paradigms and cognitive structures of human society. This makes it all the more difficult for the scientific theologian, for he has to struggle not only with the prevailing structures of social consciousness in secular life, but with the inertial force of popular religion in the Church which threatens to smother and overwhelm any genuine advance in theological understanding. Hence it is all the more necessary for the theologian, not only to question the culturally and historically conditioned habits of mind in Church and society, but to try to elevate the level of the Church's understanding above that of contemporary social consciousness, so that the Church itself may play a similar role in the world. That is surely part of the essential mission of the Church, so to live and think within human society that it enables people individually and corporately to live and think in modes of being and thought that are open to change and advance by way of response to given and transcendent realities in the world, and to the transcendent Reality of God himself over the world who is its creative ground and all-sufficient reason.

Our discussion of the social coefficient of knowledge has taken a rather spiral course in which our thought has moved constantly back and forth between the distinctive social coefficient of theological knowledge and the general coefficient of our knowledge of the intelligible world around us. They are inevitably interconnected, for the former intersects the latter from above, and the latter interpenetrates the former from below. Our particular purpose has been to clarify the status and function of the social coefficient in the tacit sub-structure of theological knowledge, and to show in some measure its suggestive and formative influence in the emergence of the communal convictions with which the theologian must operate. Without them he could not gain his initial grip upon the intangible and invisible realities into which he inquires or therefore become open to the disclosure and concept-generating power of their intrinsic intelligibility. Then as he knocks at the door of truth through his scientific questioning and finds it opening to him, the truth lays hold of him in its own objective force and sets up its laws in his mind, until he apprehends it out of itself through a restructuring of his own understanding correlated and adapted to it.

Now in bringing the discussion of this chapter to a close let me lay emphasis upon four requirements that need to be met if the theologian is to allow the social coefficient of the knowledge of God as it is embodied in the life and tradition of the Church to exercise its proper function in his inquiry.

1. The theologian must take his place within the community of living experience of God, and learn to live and think within the shared hearing and understanding of God's Word in the Church, if his mind is to be tuned into the truth of God's self-revelation and be lifted up to a level of spiritual perception and theological judgment appropriate for knowledge of God. That can hardly take place without constant training in the discipline of sanctity to match the rigour of intellectual reflection. Let me quote St. Anselm: "The heart must be cleansed by faith . . . and the eyes must be enlightened . . . And we ought to become as

little children through humble obedience to the testimonies of God . . . We must live according to the Spirit . . . The more richly we are fed on those things in sacred Scripture which nourish us through obedience, the more deeply we are carried on to those things which satisfy through understanding."[25] It is only within the milieu of the Church's worship, meditation and mission, and within its corporate orientation toward the Majesty and Sublimity of God, that the theologian may grow in purity, enlightenment, wonder and devotion and be given the capacity for forming theological concepts that answer faithfully to the revealed nature of God.[26]

2. If the theologian's task is to be fulfilled he must carry out his constructive inquiries, not alone but within a scientific community of theologians, and not apart from but in close association with the world-wide scientific community devoted to the exploration and understanding of the created universe. Through the scientific community of theologians the Church indwells the world as it comes to view in our natural scientific inquiries, and contributes to it its own distinctive perspective, in the hope that the general scientific world will in its way partake of its theological orientation toward the transcendent source of all the rational order of the universe which our many sciences increasingly bring to light. Thus the priest of God and the priest of nature are brought to work and think and learn to worship God side by side in the world he has made.

Collaboration is needed between theologians and scientists if we are to respond more adequately to the problem that arises out of the divergence between looking at the universe and looking through or beyond it. As we have already seen, looking at things tends to take away their meaning, for in meaning we attend away from ourselves to things and look through them at what they point to beyond themselves. An exclusively scientific approach raises difficulties for itself, for the more it concentrates on analysing the immanent processes of the world, and the more it restricts its focus merely to what they are in themselves, the more it tends to turn its reflection in a direction in which it can be so caught up that the meaning of the universe finally eludes it, so that the results it achieves appear empty and pointless. It would seem to be encumbent upon the Church, therefore, through its theologians so to share in and indwell world-wide scientific inquiry that it can both use it as an extension of its own objective activity in building up an understanding of God's interaction with the world, and impart to it the spiritual dimension it needs if it is not only to retain but enhance the meaning of its discoveries. I believe that in this way the Church as the community correlated to the transcendent Reality of God can share with science basic convictions and fundamental ideas which can further the heuristic activity of science itself. But I also believe that it is only as the theologian and the scientist, the community of theologians and the community of scientists, cooperate in this way that the theologian himself can meet his responsibilities and fulfil his mission adequately. This calls for a closer integration between the social coefficients of theological and scientific knowledge if there are to take place a fertile exchange and adaptation of basic convictions and concepts and a sharing in explicit understanding of the created order of the universe in its contingence upon the Rationality and Freedom of God.

3. The theologian must find a place in his inquiry into the knowledge of God for *mystical theology*. He needs to cultivate the extra-logical relation to God upon which he informally relies in the development of his theoretic system but which he cannot formalise within it. It is in this respect that dogmatic theology cannot afford to neglect the function of mystical theology in sustaining the bearing of his mind ontologically upon God himself, if it is to assume the epistemological unity of form and being both in what he knows and in his knowing of it.

A helpful, and I believe a legitimate, extension of Gödel's undecidability theorem[27] might be cast in the form: no syntactical system contains its own semantics, or, no dogmatic system carries within it its own truth-reference. To be consistent no logico-deductive or logicosyntactic system, at least of sufficient richness, can be complete, but at significant points must be open for reference beyond itself to a wider and higher system. We may also claim that no social coefficient of knowledge, no infra-structure of connections in community life, and certainly no axiomatic system of ideas can have ultimate meaning if it is closed in upon itself. Any meaningful rational system must have indeterminate areas where its formalisations break off and retain their consistency only through controlling organisation from a higher frame of reference. This brings home to us the fact that no Church as such can generate its own source of meaning but may be full of meaning in so far as it is constituted and governed by ordering principles from a transcendent ground in God. In the same vein it must be accepted that no Church dogmatics can be elaborated into a consistent and significant body of doctrinal truths through the organisation of their conceptual interrelations on one and the same epistemological level, for they require meta-theological reference to a higher level in order to be ontologically significant as well as theoretically consistent. That is to say, in the articulation and formulation of its content dogmatic theology must reckon with boundary conditions over which it has no control, basic propositions which are not decidable or logically accountable within the dogmatic system but which derive their intelligibility and force through an informal, intuitive apprehension of God who is infinitely greater than we can comprehend within the bounds of our conceptual patterns and dogmas.

It is in this light that we may appreciate the claim that mystical theology, allied to worship and meditation in the Church and the informal but objective orientation of its social coefficient toward the Transcendence of God, plays an essential role in dogmatic theology. Mystical theology functions as a constant corrective on the margins of our conceptualising and formalising operations by sustaining the non-formalisable, intuitive relation to God and by restraining the systematic impulse of dogmatic theology.[28] It takes our human forms of thought and speech, which bear the imprint of the rational structures of space and time through which God has made himself known to us, and relates them back to the ineffable Being and spiritual Reality of God, in such a way that it not only reminds us of their finite limitations but keeps them ever open toward the inexhaustible Mystery of God's Love. The cultivation of mystical theology within the experience and thought of the Church cannot but have an elevating effect upon the social coefficient of knowledge of God embedded in it. Somehow, as we have seen, that coefficient tends to develop a momentum and force of its own in virtue of which it exercises a paradigmatic control over the mind of the Church, by bracketing it within the historically conditioned parameters of the society, culture and language, in which it has taken root. However, through the transforming impact upon it of mystical theology, as well as of the on-going worship and meditation of the Church, the social coefficient can be given an open-textured disposition within which people are prompted to rise above the fixed patterns of social and religious consciousness, to be transported in wonder, and in wonder to think ahead of concept and word, and reach an apprehension of God that outstrips what can be formalised.

Mystical theology has the effect of concentrating the focus of the mind upon the ultimate ground in God which is the creative source of all our knowing and conceiving of him, but in order to be open and free for ever-deepening communion with God through his self-revelation, we must learn to *forget*, which is the other side of the heuristic outreach to what is new and hitherto unknown. The art of "unknowing" (*agnōsia*) is far from being easy.[29] To apprehend the new, old ways of apprehension must be left behind, for the new cannot be known through assimilating it to what we already know. Steady spiritual reconstruction of our individual and corporate consciousness in the Church is needed if we are to escape from the rationalisations and stereotyped modalities of the past, but it is just there that mystical theology through its enlightening marginal control of theological reflection and dogmatic formulation can fulfil its true service in our knowledge of God. Just as the Church must bring its understanding of the ways and works of God to bear upon human society, interiorising it within the general social coefficient of knowledge if the depth of meaning in the world is to be restored, so theologians must allow mystical theology to indwell dogmatic theology until it becomes interiorised within its social coefficient of knowledge, if dogmatic theology is to have the objective depth in its correlation with divine revelation which it needs if it is to fulfil its task in a godly way worthy of the Being and Nature of God.

4. The theologian cannot afford to neglect the delicate and refined slant that may be given to his inquiry through religious art or be given to the affected modification of the social coefficient of his knowledge of God upon which his inquiry tacitly relies. At the beginning of this chapter we reflected on the distinctive kind of form with which we are concerned in art, in contrast to mathematical and logical form, that is, form which through its suggestive force lifts up the human spirit to apprehend and appreciate profound aspects of reality which may not be analysed or conceptualised but which may be enjoyed for their own sake. Great works of art particularly have the power to set the mind free to reach beyond its own conceptions, to put the imaginative outreach of the mind in touch with the reality hidden behind the veil of sense and time, and creatively to call forth a capacity to appreciate them greater than would have been possible otherwise. In order to do that, however, works of art must respond to the elusive intimations of reality with such a fulness of semantic intentionality that they indicate more than can be imaginatively suggested or depicted at any time. They open out to us a surprising depth of undefined meaning which is not exhausted by the specific interpretations that are put upon it.

It is understandable that, when great art of this kind is closely associated with the adoration and enjoyment of the transcendent Beauty of God, it can become a unique partner to mystical theology in the service of the Church's worship and praise in response to what God discloses to us of his own Mystery. The place and function of Byzantine art in this respect is particularly impressive, both in relation to the social coefficient of knowledge embodied in the life and tradition of the Church, and in relation to the regular celebration of the divine liturgy and the meditation upon the Gospel. The art-form that is cultivated serves the adoration and enjoyment of God in his ineffable sublimity, and at the same time ministers to the communion of the people in the incarnational self-revelation and condescension of God in Jesus Christ. This is evident in the objective perspective of knowledge of God as it is kept wide open to the transcendent Reality of God by the artistic device in which the lines of perspective which naturally converge at a finite point are made trans-naturally to diverge into infinity. But it is also evident in the "mystical" way in which the icons are allowed to fulfil their role in the worship and meditation of the Church, not through a mimetic relation but through imageless reference to the invisible realities they intend. Thus they both guard the Mystery of God from irreverent intrusion and keep the minds of all who draw near to listen and understand the Word of God incessantly open to divine revelation. The powerful imprint that this leaves upon the social coefficient of the Church's knowledge of God makes that coefficient the matrix in which there emerge the undefined concepts which underlie and informally regulate the dogmas of the Church.[30]

NOTES

1. C. F. A. Pantin, *The Relations between the Sciences*, edit. by A. M. Pantin and W. H. Thorpe, Cambridge, 1968, pp. 100ff.
2. C. F. von Weizsäcker, *The World View of Physics*, London, 1952, p. 132.
3. M. Polanyi, *Personal Knowledge*, p. 151.
4. Cf. M. Polanyi, "What is a Painting?", *The American Scholar*, vol. 39, no. 4, 1970, p. 664f. See also *Meaning* (with H. Prosch), Chicago, 1975, pp. 82ff.
5. Cf. D. B. Harned, *Images for Self-Recognition*, New York, 1977, p. xiiif, who speaks of natural theology pursued in the light of revelation as "a confessional affair". Harned's works show a

refreshing absence of the dualism between the natural and the supernatural. See also *Grace and the Common Life*, Patiala, 1970; and *Faith and Virtue*, Philadelphia, 1973.
6. Cf. A. Einstein, *The World as I See It*, p. 174; *Out of My Later Years*, p. 60, and "Autobiographical Notes", P. A. Schilpp, *op. cit.*, p. 13.
7. On this point we still have much to learn from Athanasius. See *Theology in Reconciliation*, London, 1975, p. 237f.
8. "Readiness", *Bereitschaft*, is the expression used very effectively by Karl Barth, *Church Dogmatics*, vol. 2.1, Edinburgh, 1957, pp. 63-128, of God, and pp. 128-178, of man. The way in which he integrates "the readiness of man" into the prior and all-embracing "readiness of God", corresponds to the way in which a proper natural theology is subsumed within revealed theology.
9. St. Anselm, *Proslogion*, F. S. Schmitt, *op. cit.*, pp. 97ff. See K. Barth's exposition, *Anselm: Fides Quaerens Intellectum*, London, 1960, pp. 73ff.
10. See Karl Barth, *Church Dogmatics*, 1.1, Edinburgh, 1975, pp. 450ff. The thought here is evidently dependent on St. Basil, *De Spiritu Sancto*, 16. Cf. *Theology in Reconstruction*, London, 1965, pp. 220ff.
11. Karl Marx, *A Contribution to a Critique of Political Economy*, tr. by S. W. Ryazanskaya, and edit. by M. Dobb, New York, 1970, p. 20f.
12. Cf. H. B. Acton, *The Illusion of an Epoch*, London, 1955, pp. 133ff; and L. Kolakowski, *Main Currents of Marxism*, Oxford, 1978, pp. 335ff.
13. J. H. Walgrave, *Persons and Society, A Christian View*, Pittsburg, 1975, pp. 107f.
14. M. Polanyi, *The Logic of Liberty*, London, 1951; *Science, Faith and Society*, Chicago, 1964; *Knowing and Being*, London, 1969, Chs. 2, 4, 5; *Scientific Thought and Social Reality*, New York, 1974. And cf. my essay "The Open Universe and the Free Society", *Ethics in Science and Medicine*, vol. 6, pp. 145-153.
15. M. Polanyi, *The Logic of Liberty*, p. 28.
16. This is the point of Kierkegaard's much misunderstood insistence that "truth is subjectivity", where "subjectivity" should be understood in the sense of "subject-being". See *Concluding Unscientific Postscript*, pp. 169-224; and *Training in Christianity*, tr. by W. Lowrie, London and New York, 1941, pp. 200ff. Heidegger used the term *Subjektität (nicht Subjektivität)*, "subjecticity" not "subjectivity", *The Question of Being*, tr. by W. Kluback and J. T. Wilde, London, 1959, pp. 54-55.
17. Cf. Michael Polanyi's concept of the community of science as "A Society of Explorers", *The Tacit Dimension*, London, 1967, ch. 3, pp. 53ff. See Richard Gelwick, *The Way of Discovery. An Introduction to the Thought of Michael Polanyi*, New York, 1977, ch. 5, pp. 111ff.

18. Cf. A. Einstein, *The Way I See It*, p. 140; and M. Polanyi, *The Logic of Liberty*, p. 40.
19. We must also take into account the subjective states of others which obstruct their sharing in objective knowledge. Hence we cannot take as a working criterion what can be made understandable to others in the prevailing outlook.
20. M. Polanyi, *Personal Knowledge*, chs. 4 & 5; and *The Tacit Dimension, passim*, etc.
21. Cf. M. Heidegger, *Vorträge und Aufsätze*, Tübingen, 1959, or *Unterwegs zur Sprache*, Tübingen, 1960, etc.
22. It is in this connection that one is to appreciate what David Hume called "natural beliefs" or "propensions" which relate to the sensitive rather than the cognitive aspect of our nature, but which exercise a directive force in our cognitional activities, *A Treatise of Human Nature*, Oxford, 1967, edit. by L. A. Selby-Bigg, I.iv.1f, pp. 180ff. Cf. also Popper's idea regarding our propensity to expect regularities to which we are genetically or psychologically disposed even before our actual observations, *Conjectures and Refutations*, London, 1963, pp. 46ff.
23. See especially, M. Polanyi, *Science, Faith and Society*, and *Scientific Thought and Social Reality*. "Neither the Marxist's nor the Fascist's theory of man and society admits of common ground for argument between their adherents and the believer in transcendent reality." *Science, Faith and Society*, p. 81.
24. See *Belief in Science and in Christian Life. The Relevance of Michael Polanyi's Thought for Christian Faith and Life*, edit. by T. F. Torrance, Edinburgh, 1980, pp. 146f.
25. St. Anselm, *Epistola de incarnatione*, 1, F. S. Schmitt, *op. cit.*, vol. II, p. 8. Cf. *Anselm of Canterbury. Trinity, Incarnation, and Redemption, Theological Treatises*, edit. and trans. by J. Hopkins and H. W. Richardson, New York, 1970, p. 9.
26. For a keen perception of the heuristic property of worship, see John M'Cleod Campbell, *Christ the Bread of Life*, 2nd edit., 1869, pp. 66, 158.
27. K. Gödel, *On Formally Undecidable Propositions of Principia Mathematica and Related Systems*, tr. by B. Meltzer, Edinburgh, 1962.
28. This is the role of what Greek theology speaks of as the "apophatic" factor in our knowledge of God. Cf. John of Damascus, *De Fide Orthodoxa*, I.2, 4.
29. Pseudo-Dionysius, *Mystica theologia*, 1–5; *De divinis nominibus*, 1 & 7.
30. There is little doubt that Christian music, as it developed from early Byzantine and Gregorian sources through polyphony to counterpoint, has had much the same effect — e.g. in the bearing of Bach upon Lutheran theology. Cf. Karl Popper's remarks about two

kinds of music, "objective" and "subjective", as exhibited by Bach and Beethoven; but also about the way in which contrapuntal music provides a frame of coordinates for exploring the order of the unknown, while opening up the field for innumerable new possibilities. *Unended Quest. An Intellectual Autobiography*, London, 1976, pp. 55f, 6off, 68f. Reprinted from P. A. Schilpp, *The Philosophy of Karl Popper*, Illinois, 1974.

Chapter 5

THE STRATIFICATION OF TRUTH

THROUGHOUT these explorations in the philosophy of theological science I have tried to keep before me the conception of science advocated by Albert Einstein, both as a constant check on my understanding of what science is about, and as a foil to my own thought. I have found particularly enlightening the critical and realist epistemology that he had to develop in the course of his scientific work in order to carry thought beyond the point it had reached at the time. This was an epistemology of the kind that arises on the ground of actual knowledge. It developed along with magnificent achievement in the advance of human knowledge of the universe, and continues to be clarified in integration with it as well as backed up by it. It is an epistemology, however, that cuts rather sharply through idealist and positivist assumptions alike and results, as we have already seen, in the restoration of *ontology* in the proper sense, which those philosophies and philosophies of science that go back to Hume, Comte, and Mach find rather disconcerting, to say the least. Through the theory of relativity, and the logical reconstruction of the basis of physics which it demanded, Einstein showed that our essential scientific concepts are grounded in the invariant though dynamic structure of nature, that is, in "the physical relatedness of the physical object of knowledge", as Northrop succinctly expressed it; but Einstein also showed that "they are concepts of a kind fundamentally different from the nominalist particulars which denote data empirically given".[1] That had the effect of undermining the various kinds of analytical and linguistic philosophy which have emanated from the so-called "Vienna Circle", namely, the group of logical positivists operating from Vienna before and after the Second World War, characterised by a strong antimetaphysical bias against ontology and theology.² In contrast, as F. S. C. Northrop has shown, Einstein's epistemological theory of physical science is not a dogmatic selection from one among many possible theories but is rather one grounded in distinctions required by scientific evidence itself.³ I have found myself forced to accept this epistemology of *critical realism*, but it has challenged me to rethink the whole question of *being*, as no doubt the foregoing chapters have already made clear.

As a human being, however, as well as a theologian, I have my own problem with Einstein's thought, not that I want to object to it, as far as it goes. In any case I am not competent to do that, except perhaps on philosophical grounds, and there I find myself basically convinced by him. The point at which I am challenged by Einstein is precisely that at which he insisted in his famous essay on *Physics and Reality* that "nothing can be said concerning the manner in which the concepts are to be made and connected, and how we are to coordinate them to the experiences".⁴ I entirely agree with Einstein's refutation of Kantian and Machian ways of correlating the empirical and theoretical components in scientific knowledge, but if the scientist is a *personal being* and not just a mathematical machine, then I believe we must have rather more to say at this point. Of all our contemporaries no one has done more to advance our understanding here than Michael Polanyi, to whom I was not a little indebted in the last chapter, that is, to Polanyi's discussion of the tacit or inarticulate dimension in human thought as the area in which our spontaneous and creative thinking is at work in intimate and intuitive contact with reality. In the last chapter I tried to develop this in terms of what I called the social coefficient of knowledge, in the course of which I indicated that we are forced to take full account of *subject-being*, the being of the human knower in his active contact with what he knows, and the being of what is known as that which has the power of being open to knowledge. Whenever we operate, as we have been tempted to do regularly in post-Cartesian thought, with a subject/object relation in which the object is regarded as standing opposed to the subject,[5] and therefore with an *impersonal model of thought*, we become trapped in detached, objectivist relations to what is other than ourselves. Thus the very model of thought which we use inevitably tends to exclude *the place of personal agency* in our knowing and in the nature of what we seek to know. This is why in dualist modes of thought it is almost impossible to take seriously any understanding of God as personal active Agent in the universe. On the other hand, this post-Cartesian way of thinking collapses with the collapse of the impersonal objectivist model of thought and the cosmological dualism and mechanistic conception of the universe within which it seemed to be demanded. With the rejection of dualism in its epistemological and cosmological forms, and with the establishment of ontology in the full and proper sense, the whole situation is altered. The place of personal being in the agency of knowledge must be thought out afresh, together with the axiomatic exclusion of personal being and agency in the subject-matter of knowledge. A reassessment of our operative models of thought along these lines is certainly compelling for what we call the human and social sciences, and is, needless to say, of intense concern for theological science, but it cannot rationally be regarded as irrelevant for physical and natural science without further attention to the kind of intelligibility or rational order found manifesting itself in nature. In fact it appears to be increasingly clear in field after field of scientific research that the natural processes of expansion and evolution toward ever higher and richer forms of order embody elements of rational purpose which rule out explanation merely in terms of blind chance and necessity.[6]

All this this does not of course imply some sort of lapse into "personalism", for that would be little more than a bizarre contra-position to objectivism, operating within the same radical dualism between subject and object. At this point I am unwilling to follow Michael Polanyi, at least in certain peripheral passages of his works, in taking over as much as he does from existentialist and phenomenological thinkers, for they are still tied up with the radical disjunctions which we have had to reject in pure science, and which Christian theology rejects in its doctrines of creation and incarnation. Nor must it be taken to imply in the slightest any mitigation in the relentless objectivity of scientific thought in which we take great precautions not to obtrude our own subjective states and conditions, far less our uncritical preconceptions, into the object or field of inquiry, for properly understood, it is *personal being that is the bearer of objectivity*. What we mean by *person* and *the personal* must be reserved for the next chapter. But what we are concerned with now is the bearing of this approach upon our understanding of *truth, truth as being*, both in the human knower and in what he knows, truth as subject-being and truth as object-being, and its implication not only for the complex structures of scientific knowledge but for the whole realm of human thought and statement. I hope that the significance of the title which I have given to this chapter, "The Stratification of Truth", will become apparent as we proceed.

In order to take our bearings once again from Einstein's conception of science, let me recall several of his statements, summarise or paraphrase several others, and then offer some comments upon them.

"Physics is an attempt conceptually to grasp reality as it is thought independently of its being observed."[7] Science of any kind is possible only in so far as it rests upon "a faith in the simplicity, i.e. the intelligibility, of nature."[8] Hence what the scientist tries to do is to find the simplest possible set of concepts and their inter-connections through which he can achieve as far as is possible a complete and unitary penetration into things in such a way as to grasp them as they are in themselves in their own natural coherent structures. "Our experience hitherto justifies us in believing that nature is the realisation of the simplest conceivable mathematical ideas ... In the limited nature of the mathematically existent simple fields and the simple equations possible between them lies the theorist's hope of grasping the real in all its depth."⁹ "Behind the efforts of the investigator there lurks a stronger, more mysterious drive: it is existence and reality that one wishes to comprehend."¹⁰ The scientist is activated by a wonder and awe before the mysterious comprehensibility of the universe which is yet finally beyond his grasp.¹¹ This wonder and awe are sustained by religion. "His religious feeling takes the form of a rapturous amazement at the harmony of natural law, which reveals an intelligence of such superiority that, compared with it, all the systematic thinking and acting of human beings is an utterly insignificant reflection."¹² "By way of understanding he achieves a far-reaching emancipation from the shackles of personal hopes and desires, and thereby attains that humble attitude of mind toward the grandeur of reason incarnate in existence, and which, in its profoundest depths, is inaccessible to man. This attitude, however, appears to me to be religious, in the highest sense of the word."¹³

It is of course physical science that Einstein had primarily in view in making these statements but they present two ideas which are of particular significance for what we have in view.

1. The ultimate reference of scientific concepts and statements is to be taken back to the mysterious reality of the intelligible universe. The being of the universe is the final ground for all our concepts and statements about it.

2. The being of the universe, reality, has an ineffable character. It manifests such an infinite depth of comprehensibility that, no matter how far we manage to penetrate into it in intelligible ways appropriate to it, "our notion of physical reality can never be final", let alone anything we might think of the reasons for the existence of this state of affairs.¹⁴

So far as physics itself is concerned, and indeed so far as Einstein's own theory of relativity is concerned, this means that the more profoundly we penetrate into the ultimate invariances in the space-time structures of the universe, we reach objectivity in our basic description of the universe only so far as relativity is conferred upon the domain of our immediate observations.[15] It is precisely because our thought finally comes to rest upon the objective and invariant structure of being itself, that all our notions of it are thereby relativised. This means that knowledge is gained not in the flat, as it were, by reading it off the surface of things, but in a multi-dimensional way in which we grapple with a range of intelligible structures that spread out far beyond us. In our theoretic constructions we rise through level after level of organised concepts and statements to their ultimate ontological ground, for our concepts and statements are true only as they rest in the last resort upon being itself. Yet in so far as they have their truth in that reference, they are thereby revealed by the inexhaustible intelligibility of being to be inadequate and relative in themselves.

Let us reflect further about what it means to know being. "Everything known is known as being or as some particular form or mode of being."[16] All our knowledge in this or that science is not simply knowledge of a special field of experience, of a particular set of existents, or of some complex of relations, but in all such cases knowledge of things or events that partake of being. Hence every concept we have of things carries with it an epistemic relation to the being of beings. That is why, as Duns Scotus used to claim, the primary natural object of the human intellect is not the so-called essence or quiddity of a thing abstracted from its actual existence, but being (*ens est primum obiectum intellectus nostri*).[17] Nor is it even this or that being, but being as such (*primum obiectum intellectus nostri naturale est ens in quantum ens*).[18] In particular beings being presents to us aspects of itself for our knowledge and as such makes itself accessible to us in such a way that it admits of signification, intention, description, and so on, in its objective reality independent of us. At the same time, however, in its objective reality independent of us, it remains something which we cannot fully master: it sets boundaries against our knowing of it by the very nature of its objective reality.[19] Even when being lays itself open to our knowing it holds back something from what it allows to be disclosed to us, and paradoxically manifests its independent reality all the more powerfully by refusing to be completely netted by our concepts and declining to appear according to prediction. That is why when we seek to know anything to match our understanding of it with its nature as it becomes manifest to us, we know that it exceeds our knowing of it. Hence when we speak of it, we speak truly of it when we indicate more than we can specify, for there is far more to be disclosed than we can reach or express at any time.

Mention must also be made of a cognate feature that characterises all genuine knowledge, whether it be knowledge of some limited field of being or knowledge of the universe as a whole, namely, an inner ratio between objectivity and revisability. In so far as our knowledge is objectively grounded upon reality independent of ourselves, the concepts and statements we employ in expressing knowledge function properly by referring to that reality away from themselves, even while taking shape under the compelling claims of its inherent intelligibility. But for that very reason they are continually open to revision in the light of further disclosures of that reality which may come about. In objective knowledge of this kind there come to light invariant elements which govern our basic conceptions and affirmations of the reality concerned, but in so far as that reality exceeds our capacity to master it or resists encapsulation in our forms of thought and speech, there inevitably arise tentative variable representations of it which are to be regarded as having their truth not in themselves but in that to which they refer beyond themselves. It is only in so far as we can operate with "constants" or "invariants" in our basic apprehension of some reality that we can develop inquiry into positive knowledge and advance further by building upon foundations thus securely laid. The material content of our knowledge, what we apprehend, the objective referent, does not change, but because it is objective it transcends and relativises the conceptual and linguistic forms we develop in advancing our knowledge of it. The intelligibility inherent in the reality apprehended is unchanging, but the rational forms in which we seek to articulate our apprehension of it are variable, and must be variable in that they are inadequate to the inexhaustible nature and invariant reality of being. Their openness to revision is measured only by the depth of their objective reference.

Now all this applies not with less but with greater force to our knowledge of God. Here, of course, we have to reckon with a considerable difference between the kind of knowledge that obtains in physical science, for the created universe does not disclose or declare itself to us as God does — otherwise it would not be the creaturely or contingent reality that it is. The universe does reveal itself to our inquiries in its own limited reality, in correspondingly limited ways, but it is quite unable to explain itself or to yield any final account of the fact of its astonishing intelligibility, and so at these limits the universe by its finite nature simply turns a blank face to our questions.[20] In contrast, God opens himself to us and informs us of himself in a way that no created being can. Even though he retains behind a veil of ineffability the infinite mystery of his uncreated Being, he nevertheless unveils himself to us as the transcendent Source and sustaining Ground of all created being and created intelligibility, and therefore of all our knowing of him as well as of the universe he has made.

Moreover, the Being of God is made known to us as Subject-being, not just as Object-being over against us. As Subject-being he is the Creator and Ground of all other subject-beings, who sustains them in relation to himself as personal rational agents enabled to have communion with him. That is to say, God interacts personally and intelligibly with us and communicates himself to us in such a personalising or person-constituting way that he establishes relations of intimate reciprocity between us and himself, within which our knowing of God becomes interlocked with God's knowing of us. In fact our knowledge of God thus mediated is allowed to share in God's knowledge of himself. An ellipse of knowing, so to speak, is set up within which God's uncreated Intelligibility and our creaturely intelligibility, God's self-witness and our human understanding, are correlated, so that there arises among us within the conditions of our earthly and temporal existence authentic knowledge of God in which God's self-revealing is met by human acknowledgment and reception, and in such a way that our knowing of him, however inadequate, is made to repose ultimately on the free creative ground of God's own Subject-being.

Nevertheless when all this is admitted it still remains the case that God confronts and interacts with us as he who is utterly transcendent over all our knowing of him, infinitely inexhaustible in the Truth and Intelligibility of his own eternal Being. As such the Reality of God ever remains the Source of all our authentic concepts of him and the unchanging Ground of all our faithful formalisations of his revelation. God himself does not change, and in his unchanging Being is open to ever deepening understanding on our part, while our forms of thought and speech in which we articulate our knowledge of him are ever open to further clarification, fuller amplification, and change. The Truth of the divine Being cannot be enclosed within the embrace of our finite conceptualisations. In that God admits of recognition and understanding on our part we may indeed grasp him in some real measure, but we cannot contain him in the forms of our grasping. We may apprehend God but we cannot comprehend him. In so far as our concepts of God derive from him and terminate upon his Being, there is much more to them than the concepts themselves, more than the formal truths of conception, for the Reality conceived transcends conceptual control. Before the Reality and Majesty of the divine Being whom we are graciously allowed to know, we know that all our knowledge of him is at a comparatively elementary level, and all our articulation or formulation of divine revelation is a relatively insignificant reflection of its Truth. The knowledge and understanding of God, however, which we are allowed to have, and which in some measure we may bring to systematic expression, are what they are in their lowly forms because, in spite of their utter inadequacy, as the human end of the ellipse of knowing established by God and maintained between us and himself, they are locked into an infinite range of truth and intelligibility grounded finally in God's own eternal Being.

The development of our knowledge of God evidently involves a multi-levelled structure in which our thought moves through various levels of concepts and statements, to the levels of created being through which God makes himself known to us in space and time, and then through them ultimately to the supreme level where God is the transcendent Source of all truth in the Truth of his own uncreated Being. Each lower level is governed by reference beyond itself to the level with which it is immediately coordinated, so that together the lower levels constitute a coherent semantic frame of reference through which we are directed to the ultimate Truth that God is in himself. Thus every lower level, in so far as it is true, must have the character of an open structure pointing us away from its own limited and relative status to its ontological ground in God who is "the norm for the truth of all beings".[21] In clarifying and deepening theological knowledge, therefore, we must learn to penetrate through the various levels of rational complexity that arise in the process of inquiry to the ultimate ground upon which they rest in the Being of God. Just as we do not think statements or even normally think thought but think things through them or by means of them, so the structures of the reason which arise in the process of gaining knowledge have to be treated as refined conceptual instruments through which we let reality shine across to us, in order that its own truth of being and inherent intelligibility can operate creatively in our understanding of it.

What are we to understand by "truth" in a context like this?

In its primordial and most basic sense, as Patristic and Mediaeval theologians used to point out, truth denotes the state of affairs that is ontologically prior to the truth of cognition or the truth of statement, that is, the *truth of being*, which is more or less synonymous with reality. For St. Augustine, for example, truth simply is "what is", or, more fully expressed, "that which manifests what is . . . and manifests it as it is".[22] The concept of truth thus enshrines at once the real being of things and the revelation of things as they are in reality. The truth of being comes to view in its own light and in its own authority, constraining us by the power of what it is to assent to it and acknowledge it for what it is in itself. St. Anselm, who developed this further in a more realist way, held truth to be the reality of things as they actually are independent of us before God, and therefore as they ought to be known and signified by us. Everything is what it actually is and not something else, and cannot according to its nature be other than it is. That is what he called its inherent "necessity" or "truth".[23] In the nature of the case, it must be discerned and known strictly in accordance with the necessity of its being what it is or in accordance with its inherent truth of being. Thus to the truth or necessity in the object of knowledge, that a thing is what it is, there must correspond a truth or necessity in knowledge, the impossibility of conceiving the object as being other than it is. Hence all our concepts and statements must be tested through a critical inquiry in which we listen in to the truth of things, yielding our minds to their ontic necessity, for in that way we are enabled to straighten out our conceptions and statements in accordance with the nature of things as they really are and so think and speak rightly of them. Only when we are able to embody in our thinking and in our discourse a rational structure of signification which reflects or participates in the inherent rationality of truth (*ratio veritatis*) have we engaged in scientific activity — then only may we have some title to speak of true knowledge.

It may help to clarify this concept of truth — with which I am in basic agreement — if we draw several distinctions.

1. In speaking of the truth of being we must distinguish between *the supreme Truth and the truth of created realities*.

God is the supreme Truth in the sense that while all other beings have their truth ultimately by reference to what he is, he alone, as the self-existent Being, is known in the light of his own Truth. The Truth of God is that he is who he is and reveals who he is as he is. He confronts and interacts with us in the freedom and majesty of his own Reality, in the prerogative of his own Truth, in the light of his own Being, in the power of his own Self-evidence. He is the ultimate Truth who cannot be recognised or known by reference to anything else, who cannot be bracketed with any other reality, and who cannot be brought under the mastery of our comprehension, but who nevertheless makes himself accessible to us for he is not limited by our littleness or weakness. We may know him truly under the power and constraint of his being what he is and what by his nature he must be for us, that is, under the light of his own divine Truth which is his divine Being coming to view and becoming in our understanding and knowledge of him what he is consistently in himself and in all his relations with us.

As such, however, God has created the universe to which we belong, giving to created beings truth of their own, so that they may be known truly only when they are known in accordance with the inner necessity or reason of their contingent beings, for they are, under God's creation, what they are and not other things. And since they cannot according to their natures be other than they are, they must be known in accordance with their ontic necessity, or their inability to be otherwise. It is because God the Creator stands behind the things he has made, that we respect God himself when we respect the nature of created beings and act toward them strictly in accordance with what they actually and contingently are. Thus created realities have their own created truth of being which we must acknowledge if we are to have right relations with them, conceiving of them and speaking of them, and indeed acting toward them, under the constraint of their inherent truth of being.

2. We must also distinguish between *truth and truthful-*

ness, the truth of being itself from the truth of relation. We owe it to the truth to be truthfully related to it. Truthfulness is an openness to the truth and a rightness of relation to it in accordance with the obligation which the very nature of the truth lays upon us. While truthfulness involves an analogical or referring relation to the truth, the truth itself always retains ontic priority, for it is what it is in its own independent reality before it is recognised by us, and what it is in its own inherent rationality is the compulsive ground for our recognition of it, assent to it, and as such the inexhaustible ground of our conceptions of it.

3. Again, we must distinguish between *the truth of being and the truth of statement*. It is impossible to picture how a picture pictures what it pictures, for that would involve the absurdity of reducing to a picture the relation between the picture and the reality pictured. No more can we state in statements how statements are related to being, without reducing the relation of statements to being entirely to statements. To attempt to do this is to resolve the truth of being into statements about it. Here we must distinguish between a true statement and the truth of statement, and maintain a true or right relation between a statement about the truth and the truth itself. It does not follow that because a statement about the truth must be distinguished from the truth of being to which it refers, that it is not objectively grounded in the truth or controlled by it. Even if we cannot argue from the statement itself to the reality which it signifies and in which it is intelligibly grounded, nevertheless as the statement directs our minds to the reality, the reality itself, which is quite independent of the statement, shows itself and thrusts itself upon us in accordance with its own nature. Whenever a statement serves the truth of being like that, it is a true statement, but whenever it obstructs that disclosure of the truth of being, it is a false statement.[24] Hence in scientific thinking we do our best to match the signifying of our statements rightly to the power of being to be signified, to let it be controlled by the inherent reason of the truth of being (St. Anselm's *ratio veritatis*), and so to let it show through to us in its own

reality in such a way that we cannot rationally, truthfully, or in good conscience, decline to assent to it.

In view of this it is evident why St. Anselm in laying the foundations of scientific theology in the West insisted that we must distinguish between several levels of truth.[25]

1. There are the *two truths of statement*. When a statement is considered by itself merely as a formal part of speech, apart from the truth or falsity of any reference, and in so far as it makes verbal sense, it can be said to be "true" when it does what it ought to do. It can do this (a) by signifying what it is capable of signifying, i.e., by fulfilling its syntactical functions in a consistent and coherent set of words; and (b) by signifying in accordance with what it is made to signify, i.e., by fulfilling a semantic function in referring to a state of affairs beyond itself. When a statement does what it ought in these two ways it has, St. Anselm claims, two "truths" or "rightnesses", one which is always "natural" to it as a meaningful sentence, and one which is "accidental" to it, depending on the use to which it is put as a proposition. But of course it could make sense as a sentence and still be false by failing to state what is actually the case. If it has truth in the second way, its truth or rightness will depend on the truth of what it undertakes to signify. Properly speaking, therefore, when a statement is true it must be true in both ways, that is, when a statement that is grammatically meaningful as a sentence is employed in a way that it refers rightly to a state of affairs in things beyond itself.

2. Bound up with the truth of statement is the *truth of signification*, which is found when both poles of signification have their place, with a rightness in the signifying statement and a rightness in its relation to the thing signified. But since this depends on the nature of the thing signified, the truth of signification depends on the truth of the thing signified. The truth of signification, then, is not located in the statement itself, for since it signifies what is the case, it depends for its truth on the truth of what it signifies independent of it. It signifies rightly when it signifies what it ought by signifying what is in accordance

with the facts. That a thing is what it is and not another thing demands that we signify it rightly in accordance with what it is and not otherwise.

3. The truths of statement and signification presuppose the *truth of being*, or what St. Anselm calls "the truth of the essence (or existence) of things" (*veritas essentiae rerum*). The truth of a thing or of a being is its reality, what it actually is and what in accordance with its nature it ought to be. This truth or rightness of things, however, is not immanent or self-subsistent in them as such, but is in them only so far as they are truly or rightly related to their creative source in the Supreme Being who only is Self-existent. That is to say, the truth of being anywhere within the creation is a contingent truth, dependent on the Supreme Truth which God is in his own eternal Being. Whatever is, therefore, is truly in so far as it is what it is and what it ought to be there, in the Supreme Truth, for then it is as it ought to be. Thus the truth of created being has an in-built reference beyond itself to its ultimate ground in God.

4. The *Supreme Truth* is the Truth of the self-existent Being, the Truth of God, Truth in its own right without reference beyond itself. This is the ultimate Truth to which all other truths finally refer and by reference to which they are themselves truths. It is thus from the Truth in this supreme sense that there derives the universal obligation for things to be true, and for concepts and statements to be true as well. Hence whenever we speak or think of things as true, we are engaging in a transcendent reference to and beyond the truth of things to their ultimate ground and justification in God the one Source of all being and truth.

This stratification of truth, as I have called it, is of immense importance. By finding the truth of statements in their referring beyond themselves, and by finding the truth of signification in signifying that that which is, is the case, St. Anselm has shown the impropriety of reducing the truth of statement to its truth-function in discourse, for the truth of the thing signified remains when it is not signified or when it is signified falsely. On the other hand, by finding the truth of that which is, namely, the truth of being, in a transcendent reference beyond in God, St. Anselm has shown the impropriety of defining truth in terms of agreement with the laws of human understanding, since the truth of things remains even when we do not think of them. Their truth is rooted independently of us in a structure of reality that derives from God and not in a structure which we impose upon things by our knowing of them.[26] Moreover, this stratification of truth clarifies the objective depth and range of all rational and scientific inquiry, in which we seek to let our minds fall under the compulsion of the actual nature of things, and to bring to light their inner logic (once again, St. Anselm's *ratio veritatis*) independent of our perceiving of them and thus their objective invariances irrespective of any or all observers. But this stratification of truth is important also for another reason: it shows the subordination of all our conceptual and linguistic formalisations to the inherent intelligibility of reality, and invites inquiry to penetrate more and more deeply into the rational structure of reality in its profound unity and simplicity.

This way of describing the stratification of truth has its considerable merit, for it sets out the different levels of truth in their cross-level coordination with one another. Each level is found to be open to the level above it and to require that meta-level relation in order to be consistent in itself as a level on its own. Thus there becomes disclosed the organic structure of thought that characterises our apprehending of reality in all its depth. It must be recognised, however, that this is a posterior reconstruction of the way in which our knowing is coordinated with the reality known, and does not take into account the actual heuristic processes of inquiry or the complex systems of thought and statement which build up in the course of those processes as the hypothetical compounds through which we try to penetrate more and more deeply into the objective structures of reality.

In order to understand something of the actual gradient of inquiry, however, we must consider it all from the other way round, from within the perspective of our inductive struggles to reach the truth, and thus to trace out how we build up different logico-syntactical systems at different stages in the advance of knowledge. This procedure will also serve to reveal the difficulties and problems with which we have constantly to cope in the verification and clarification of our knowing.

Let us glance first at what we do in natural science, taking our cue once more from Einstein, and remembering his point that science is after all a refinement and extension of our everyday thinking.[27]

We start with our ordinary experience in which we operate already with some sort of order in our thought which is essential for our understanding of the world around us and for rational behaviour within it. We assume that the world is intelligible and accessible to rational knowledge, for otherwise our thoughts and experiences could not be coordinated. And so we operate on the assumption that by means of thought we can understand in some real measure the relations between events and grasp their orderly sequence and consistent structure.

What we are concerned with in science, however, is to deepen our grasp of that orderly structure. We select a few basic concepts in our experience and apprehension of the world, try to work out their interconnections, and organise them into a coherent system through which like a lens we can gain a more accurate picture of the hidden patterns and coherences embedded in the world. We carry out this activity in other fields of investigation and try to connect together the various structures we discover latent in them, thus widening and at the same time unifying the progress of our science. But all the time we penetrate more and more deeply through different systems and levels of rational complexity that arise in the course of our inquiries as nature manifests itself to us and even discloses to us objective structures that are inherently non-observable but which constitute, so to speak, the invariant back-side of reality.[28]

In the course of these inquiries, as we have already noted

in an earlier chapter, we operate with distinctive modes of rationality in the light of the nature of the world as it becomes progressively disclosed to us and as far as we are able to conceptualise and formalise it. Thus the logico-deductive systems we build are related to different conceptions of rational order which inevitably undergo modification and change the more fully we uncover the astonishing intelligibility everywhere immanent in the universe. Thus there are found in the history of Western thought decisive turning points in a scientific outlook upon the cosmos, from the Ptolemaic to the Copernican cosmology, with a corresponding change in mathematical instrumentality, as with the discovery and employment of "fluxions" in Newton's determination of the differential laws of the universe as a system of bodies in motion; and then from the Newtonian to a very different cosmology, beginning with Clerk Maxwell's application of partial differential equations in an explanatory account of the mathematical properties and structure of continuous fields of force, and culminating in Einstein's "relativistic" outlook, with a switch away from a mechanistic conception of the universe, and the development of mathematical invariances to describe the dynamic structure of the space-time metrical field which embraces and controls the whole finite but unbounded universe. At each of these stages the concepts we develop are integral to a whole cultural and rational system, so that advance can be made only with a radical shift in the basic fabric of thought together with a logical reconstruction of prior knowledge. This is not to claim that the basic concepts we reach in immediate intuitive experience of the world are changed, any more than our basic rationality alters, but that we reach a deeper and more appropriate understanding of our basic concepts and through them of the dynamic, multivariable world with which we are so marvellously and intelligibly coordinated. This is why scientific inquiry advances only as it clarifies and deepens the foundations laid in the past. Otherwise it would be little more than a very sophisticated game without any empirical application to actual existence at all.

Mutatis mutandis, much the same procedures are to be found in the development and advance of theological knowledge, but here the inductive struggles to reach the truth, or rather to let the truth have its way with us, are rather more troublesome, for they have to cope with the stubbornness and intractability of self-centred human nature at deeper levels. We should not let ourselves think that natural scientists are quite without struggles of this kind for in pursuing their objectives they need to keep free from the shackles of selfish desire and constant preoccupation with themselves, as Einstein himself used to declare, in making the point that the kind of emancipation which the great scientists have achieved in this respect is religious in the highest sense of the word.[29] In the knowledge of God, however, there is a much profounder correlation between the human knower and what he seeks to know than in natural science, and in fact what John Calvin used to call a mutual relation between knowledge of God and knowledge of ourselves.[30] For that very reason theological knowledge has to strike more deeply into human being, and into the self-centred or in-turned character of man's self-consciousness. That is to say, it has to struggle with the persistent selfishness of human nature and with the deep-seated alienation of the human mind from the Truth of God which belongs to what theologians call "original sin". If the advance of scientific knowledge can be made only with a radical shift in the basic fabric of our thought together with a logical reconstruction in the foundations of prior knowledge, real advance in theological understanding and knowledge must entail correspondingly a radical change in the basic structure of the religious self. The more rigorously theological science asks its questions the more it is forced to question its questions and reconstruct them if they are to serve inquiry into the truth. But if the questions that arise are inextricably rooted in the human self, so that in the last resort the questions we put to God are ourselves,[31] then it is we ourselves who have to be questioned down to the very roots of our being. And of course that is what we bitterly and deeply resent. Hence in the very nature of the case processes of atonement, forgiveness and reconciliation have to operate step by step with theological advance, for men and women need to be reconciled with the truth if they are to know the truth, and they cannot really know it without becoming at-one with the truth, which cannot but involve radical self-denial and conversion on our part, as Jesus made so clear when he called disciples to renounce themselves, take up their cross and follow him with a radical reorientation of their minds. Knowledge of the truth and repentance go hand in hand together.

Our immediate concern here is not so much with the individual as with the social aspect of this process. In so far as our social consciousness becomes infected with our inherent selfishness, the social coefficient of knowledge develops a lordly, prescriptive tendency, in which we project the changing patterns of our social consciousness on to the structure of what we seek to know, and thus distort and misshape it. We use it for our own purposes and make it serve our own pragmatic policies. That is to say, in the development of different levels of rational complexity, analysis reveals that the logico-syntactical systems we employ are culturally conditioned and determined, and as such they reflect the different stages and contexts of cultural development throughout the centuries. Moreover we find that we have to take into account a conditioning of theological thought and formulation by particular powerful centres of culture bound up with language and race and national ideology. Such a state of affairs became very soon manifest at the Reformation in the principle so blatantly advocated in the well-known formula, *cuius regio, eius religio*.[32] Thus cultural conditioning of thought has not only been brought sharply to light in our day through socio-historical analysis, but paradoxically has been greatly accentuated through the rise and dominance of the social sciences, and not least in their call for sociological control of our ways of life, without respect to judgments of value or of truth and falsity. Here we must also reckon with the cybernetic thrust of the technological society, not only in Marxist countries, which threatens the purity and objectivity of science. In some respects, however, socio-cultural conditioning exercises its subtle pressures more forcefully in theology.

Particularly difficult and persistent problems arise for theology through the mission of the Christian Church to communicate the Gospel to every generation in contemporary terms, to make the message understandable to people in their own language, condition and culture. This means that the knowledge of God as it is articulated in the Church tends to be geared into the spirit of the times, assimilated to the transient patterns of life and the various symbolisms of art and literature through which it expresses itself and in terms of which it develops its distinctive web of meaning. It is not difficult to see that in this way the Church is constantly tempted to communicate its message to people in terms of the residual ideas of a past and not simply a passing culture, so that it tends to build into the understanding of its people an inevitable obsolescence, when of course it ought to be doing precisely the opposite, in communicating a Gospel which calls for repentance and a radical change of mind, and a forward advance in constant renewal of its life and thought face to face with the living God.

When from this point of view we look upon the historical divisions in the Church we can see that instead of being a community of reconciliation and unity in the world the Church has tended to let the cultural divisions and the group-egoisms on which they thrive in the world cut back into its own existence with schismatic results. Fragmentation and pluralism in the Church reflect fragmentation and pluralism in our secular culture. That is doubtless why, especially when the arts and humanities everywhere suffer from a disintegration of form, the ecumenical movement finds it very hard going today.

It is from this perspective also that we are to discern the reasons for the undue multiplicity of belief and doctrine in the Church. Many beliefs and doctrines evidently derive from the unending questions posed in different cultural contexts, systems and ages, for the answers given to them reflect particular sets of circumstances and the transient frames of reference they threw up. Thus they must be regarded as often little more than forms of applied theology with only *ad hoc* temporary relevance to the variable coefficients of social development rather than to the deep invariances in our human relation to God. That is why, when these time-conditioned formulations of belief and doctrine are perpetuated beyond their original sphere of origin and relevance, they become meaningless factors in our thought, obsolete clutter ministering to confusion and disorientation in the Church. Moreover again and again locally and temporally derived doctrines get pushed into the centre and through controversy come to occupy a far larger share of importance than they deserve, so that it becomes rather difficult for the untrained theologian to distinguish between what is basic and essential and what is peripheral and even perhaps quite irrelevant.

In view of this state of affairs it is extremely important for us today to recover and operate with the realisation that truth is to be apprehended in and through different levels of complexity, and that a very proper stratified structure develops in our understanding of it. What we need is a rigorously scientific examination of the whole corpus of theological knowledge in which we try to isolate the core of basic and central theological concepts and relations, as few in number as possible, distinguishing them from derived notions of an intermediary and secondary nature, in order to grasp something of its inner coherence and unity, and then use it as an instrument with which to comb through the whole corpus of accumulated beliefs and doctrines in the service of clarification and simplification. This means that, working in the whole history of theological inquiry and development, we must examine the different levels of symbolic and systematic formalisation, and try to penetrate through them until we can grasp as far as possible the basic structure of our knowledge of God, and then in the light of that employ something like "Occam's razor" to slice away all unnecessary accretions and layers in the elaborated machinery of our thought.

Without doubt the rehabilitation of a realist approach to knowledge which gives priority to the truth of being over truths of signification and statement opens the way for considerable clarification and simplification by making them point beyond themselves to a unifying ontological ground. This is bound to undermine a nominalist approach to knowledge in which concepts and statements about the truth are identified with truth itself, which inevitably leads to a steady proliferation of particular truths claiming independent status for themselves and calling for formal, logical interconnection in detachment from material content and apart from any control through objective reference beyond themselves. In the actual history of dogmatic theology, and not least in the tradition of the Roman Catholic Church, the multiplication of theological concepts and doctrinal propositions demanding assent on the ground of their alleged identity with truth, sooner or later calls for belief in entities to which it is felt they must correspond, so that a damaging pluralism becomes introduced into the material content of divine revelation. Hence in view of the rise and dominance of a great host of abstract concepts, distinctions and definitions, too often objectified and given pseudo-ontological status in the general body of received or textbook theology, it is hardly surprising that Pope John XXIII at the beginning of the Second Vatican Council should warn the assembled fathers to distinguish the substance of the faith from expressions or formulations of it.

Further, the stratified structure which we have been considering (in which signification and statement are found not to have their truth in themselves but in truths of being to which they refer and from which they are to be discriminated, and truths of being for their part are found not to be self-sufficient but to bear contingently on the one Supreme Truth of God) shows us that theological understanding and doctrinal formulation are properly grounded in God's unique self-communication to us in Jesus Christ his incarnate Word through whom we have access to the Father in one Spirit. Under the constraint of the semantic focus constituted by this "hierarchical" coordination of different levels of truth, theological knowledge is progressively concentrated upon the objective centre and content of God's self-revelation in Jesus Christ which is identical with the Truth which God eternally is in the mystery of his own triune Subject-being. That is to say, to give ontological and controlling priority to the oneness of God's self-revelation through the Son and in the Spirit, the one comprehensive Truth from which all other truths derive and to which they all finally refer, cannot but have the economic effect of purifying and simplifying theological truths, if only by disclosing that, while relatively few have really central significance, others have but an off-centre or merely peripheral significance. This is after all the way in which scientific theology at its best has always tried to proceed, although it has constantly to struggle with a traditionalist theology that is deeply affected by the ground swell of popular piety or culturally conditioned patterns of religious self-consciousness in the Church, and is not infrequently overlaid by sets of legalistic ideas and prejudiced formulae generated through theological controversy and given authoritative status through the self-justification of the Church.

It may be worth glancing again at what happens in scientific inquiry, which has to cope with similar problems and find similar solutions. This is apparent in the way in which new ideas tend to be resisted under the inertial drag of so-called "community paradigms",[33] but also in the need for scientific advance to engage in some sort of conceptual surgery to cut away unnecessary notions or dislodge rejected theories whose life is anachronistically prolonged in text-book science. In classical physics we may recall the place given by Newton to abstract notions of time and space which he invested with a causal absoluteness to provide him with the unchanging geometrical magnitudes he needed to give scientific expression for unvarying laws of nature. Necessary as they may well have been at the time, and perhaps inevitable at that stage in the advance of physics, Newton's concepts of absolute time and space had to be abandoned along with the mechanistic interpretation of nature which they imposed. This does not mean, of course, that Newtonian physics was completely invalidated but rather that it had to be reconstructed as a limiting case of relativistic physics.[34] It may also be argued that the concept of ether was a necessary conceptual tool at a certain stage in the development of our understanding of the behaviour of nature, when it functioned as an immobile substratum for electromagnetic phenomena and as a means of connecting up concepts and relations in accordance with Newton's laws of motion. However, when it became clear as a result of Clerk Maxwell's equations that electromagnetic waves including light waves propagate themselves in free space conditions and that fields of radiation are not states of a medium but are independent realities that are bound to no medium, the notion of ether fell away quite naturally.[35]

Examples like this of what happens again and again in the course of scientific inquiry can easily be multiplied, but enough has been said to indicate that in the advance of human thought we develop theoretic constructs and connections which may well be necessary within certain limits, but which are later found to be no more than intermediate devices, ladders by which we climb from one level to another but which can be, and must be, kicked away when they have done their job. In this process we discover that we operate with a core of basic concepts and relations closely bound up with our experience, and that there are other concepts which we develop in the form of theorems to connect our basic concepts together but which are of a more peripheral or merely functional significnce.[36] As they enable us to get a deeper and simpler grasp of the inner structure of things, they themselves are relativised and are progressively left behind. In this way we are able to organise our knowledge into various levels of thought with fewer and fewer fundamental concepts. As the levels coordinated in this stratified way become narrower and narrower, they focus our thought more and more sharply upon the inherent patterns of nature enabling us to grasp them in their objective unity and simplicity as far as that is possible for us. In so doing, however, we find that the fewer or the more economical and natural our basic concepts are, the wider the range of their applicability to the universe turns out to be.

This is the kind of epistemological and constructive activity for which I am calling in theological science: such a penetrating grasp of the organic structure of our knowledge of God in its inner unifying core that we are able to discriminate what is relevant from what is irrelevant, what is central from what is merely peripheral, what is basic and constant from what is superficial and variable, and what is of permanent from what is only of transient significance.[37] Then in the light of this insight we need the courage to prune the vast growth of theological propositions and doctrines very drastically, and so bring to light the fundamental unity and simplicity of our knowledge of God. Such a unification and simplification of theological knowledge can be achieved only as we pierce through layer after layer of pseudo-interpretations, substitute symbolisms and secondary formalisations to the objective basis and inherent intelligibility of its proper ground in divine revelation. So far as Christian theology is concerned such an *a posteriori* reconstruction of theological knowledge will surely bring into open and sharp relief the controlling ground-structures of our knowledge in the incarnation, creation and the Trinity, within which the obedient humanity of Jesus Christ and his filial relation to the Father will inevitably occupy a critical and constructive place in our thought.

This will have the effect of concentrating theological attention upon the cardinal issue of the oneness in being and agency between Jesus Christ and God which was soon found by early Christian theology to be the decisive hinge upon which the whole Christian outlook upon God and the created universe turned, and given the role of a supreme regulative principle in the great Conciliar Theology from Nicaea to Chalcedon, when the classical foundations were laid upon which all Christendom has ever since rested. From that controlling centre, however, the relative significance and relevant place of other traditional doctrines will be determined and plotted as the core of essential Christian belief is applied to every aspect of our human existence and activity in the creation. The closer attention is given to the centre, the more the mind of the Church will be free from sociological constraint and secular pressure to compromise the truth, but the more applied theology moves from the centre to the periphery the more variable and flexible it will be, not least in those areas where rigid forms of ecclesiastical pronouncement regarding dogma and order have played such havoc in dividing the world-wide community of the faithful. In the great hierarchy of truths, to be absolutely related to what is of permanent and paramount importance in the centre, carries with it a requirement for us to be only relatively related to everything else. Thus the economic ordering of all belief and doctrine from a given centre in God's self-communication to mankind in the Incarnation, makes for a greater measure of openness and freedom than we may have known before.

I would also hope that an approach to the theological task along these lines would enable theologians to enter into closer dialogue with scientists in every area of human inquiry than has been possible hitherto. I do not see how this can be achieved, however, until we have theologians who in dedicated respect for their subject-matter are prepared to undertake on their own ground the same kind of strenuous and arduous intellectual activity that our physicists, mathematicians, chemists or biologists manifest in their fields. Certainly until we put our own house in order by carrying through a thorough clarification and reconstruction of our own theological foundations in such a way that their fundamental unity and simplicity are brought to light, theology will hardly be in a position adequately to interact with rigorous science in any other field, or therefore to fulfil its God-given function in the modern world.

NOTES

1. F. S. C. Northrop, in P. A. Schilpp, *Albert Einstein: Philosopher-Scientist*, pp. 397, 405.
2. Cf. the works of L. Wittgenstein, F. Waissmann, and A. J. Ayer, among others, in Britain, and the excellent account of philosophy in this vein by J. O. Urmson, *Philosophical Analysis. Its Development Between Two World Wars*, Oxford, 1956. The influence of Viennese positivism is also evident in the field of law, e.g. in the works of Hans Kelsen, *The General Theory of Law and State*, Cambridge, Mass., 1945, or H. L. A. Hart, *The Concept of Law*, Oxford, 1961.
3. F. S. C. Northrop, *op. cit.*, p. 403.
4. A. Einstein, *Out of My Later Years*, p. 61.
5. This notion of "object" (*ob-jectum*) is obviously retained in the German *Gegen-stand*, which has influenced Heidegger's analysis.
6. See, for example, recent essays by George W. Tyler, "Goal-seeking Behaviour in Queueing Systems", *Operational Research Quarterly*, vol. 27, No. 3, 1967, pp. 605–614; and "Purposeful Control", *Journal of the Operational Research Society*, vol. 29, No. 2, 1978, pp. 97–104.
7. A. Einstein, "Autobiographical Notes", in P. A. Schilpp, *op. cit.*, p. 81.
8. A. Einstein, *ibid.*, p. 63.
9. A. Einstein, *The World as I See It*, pp. 136 & 137f; *Ideas and Opinions*, pp. 274f.
10. A. Einstein, *The World as I See It*, p. 140.
11. A. Einstein, *Out of My Later Years*, pp. 30, 60.
12. A. Einstein, "The Religious Spirit of Science", *Ideas and Opinions*, p. 40. Cf. also "On Scientific Truth", *ibid.*, p. 262.
13. A. Einstein, *Ideas and Opinions*, p. 49.
14. A. Einstein, *The World As I See It*, p. 157. Cf. *Out of My Later Years*, p. 95.
15. Cf. H. Margenau, "Einstein's Conception of Reality", in P. A. Schilpp, *op. cit.*, p. 254.
16. Julian A. Hartt, *Being Known and Being Revealed*, p. 10f.
17. John Duns Scotus, *Ordinatio*, prol. n. 1; I.d.3 n. 118, 137, 171–174.
18. John Duns Scotus, *Ordinatio*, prol. n. 1.
19. See Julian A. Hartt, *op. cit.*, pp. 35ff.
20. I have borrowed this expression from Hartt, *op. cit.*, p. 36.
21. Clement of Alexandria, *Protreptikos*, 6.
22. St. Augustine, *Soliloquia*, II.5.8: *verum mihi videtur esse quod est; De vera religione*; 36.66: *intelligit eam esse veritatem, quae ostendit id quod est ... ipsa est quae illud ostendit sicut est.*
23. St. Anselm, *De veritate*, *Opera Omnia*, edit. by F. S. Schmitt, Edinburgh, 1946, vol. 1, pp. 173–199. In relating "necessity" to "truth" in this way, St. Anselm is emphasising the fact that the

truth is unable to be other than it is. No "necessitarianism" is intended. Cf. Karl Barth, *Anselm: Fides Quaerens Intellectum*, London, 1958, pp. 49f. For the following see also my essay "The Ethical Implications of Anselm's De Veritate", *Theologische Zeitschrift*, vol. 24, 1968, pp. 309–319.
24. See again, St. Augustine, *De vera religione*, 36.66.
25. St. Anselm, *De veritate, op. cit.*, pp. 177ff, 185ff, 189f.
26. Contrast Immanuel Kant's intellectual notion of truth: "For truth or illusion is not in the object, in so far as it is intuited, but in the judgment about it, in so far as it is thought." *Critique of Pure Reason*, tr. by Norman Kemp Smith, B.350, A.294.
27. A. Einstein, *Out of My Later Years*, p. 59; *Ideas and Opinions*, p. 290.
28. A. Einstein, *Out of My Later Years*, pp. 95f; and cf. Northrop, in P. A. Schilpp, *op. cit.*, pp. 396f; and V. F. Lenzen, *ibid.*, pp. 366ff, 370ff.
29. A. Einstein, *Ideas and Opinions*, pp. 40, 44, 48. At these points, however, Einstein's apparent identification of the personal with the selfish or self-centred would seem to prejudice him against the personal, so that he retreats into an impersonal model of thought which nevertheless allows of "superpersonal objects and goals", p. 45.
30. John Calvin, *Institutio*, I.1.1–3.
31. Cf. Paul Tillich, *Systematic Theology*, London, 1953, vol. 1, pp. 181ff.
32. This principle was put into force by the Religious Peace of Augusburg in 1554.
33. Cf. T. S. Kuhn, *The Structure of Scientific Revolutions*, Chicago, 1962, Ch. III.
34. Cf. A. Einstein, *The World As I See It*, p. 172: "Let no one suppose, however, that the mighty work of Newton can really be superseded by this or that other theory. His great and lucid ideas will retain their unique significance for all time as the foundation of our whole modern conceptual structure in the sphere of natural philosophy."
35. Cf. A. Einstein, "Relativity and the Ether", *The World As I See It*, pp. 193–204.
36. Cf. Einstein's account of "the stratification of the scientific system", in "Physics and Reality, Out of My Later Years", pp. 58–64.
37. A similar notion was suggested in *De oecumenismo* of the Second Vatican Council, 2.11, *Conciliorum Oecumenicorum Decreta*, edit. by J. Alberigo *et al.*, Bologna, 1973, p. 915.

Chapter 6

THE TRINITARIAN STRUCTURE OF THEOLOGY

IN following the line I have taken in several earlier chapters I would like to begin this one by referring again to Einstein's conception of science. According to him the supreme task of physical science is to arrive at those universal elementary laws from which knowledge of the universe can be built up. Through painstaking and sympathetic handling of experimental material he selects and refines his concepts, reducing them and their interconnections to as few as possible, and out of them projects the simplest possible system of thought by means of which to achieve a complete and unitary penetration of natural events. In this process he must develop a logically unified conceptual representation of physical reality, but his concern is not with any conceptual system as such so much as with the physical relatedness inherent in nature, that is, with the physical structure of reality itself. Only basic concepts that have reference to experience can be of ultimate use to him, so that they must be derived not by way of logical abstraction from experience but through intuitive apprehension of experience, for it is that empirical connection that determines the cognitive value of his system of concepts. This brings its own difficulties, however, for even when the physicist goes on to refine these concepts he is compelled to operate with words which are inseparably connected with primitive, pre-scientific concepts, while the more he achieves his aim of simplicity in the theoretic basis of his thought, the more he has to accept the fact that that theoretic basis becomes further and further removed from the realities of experience, so that it becomes harder and harder to correlate his thought with actual experience of the real world of space and time. Yet it is only in this way that he can reach a profound grasp of reality itself.[1]

The task and problems of a scientific theology are not very far different from that. As I tried to show in the last chapter, it is incumbent upon us to carry through a critical revision, unification and simplification of the whole body of theological knowledge as it has grown up in history, and when we do that our thought is carried through to the conception of the Trinity of God as the basic grammar or ground structure of Christian theology. Hence in the *doctrine* of the Trinity we have a refined model comprising a minimum of basic concepts immediately derived from divine revelation and our intuitive apprehension of God in his saving activity in history, together with a minimum of secondary concepts or relations of thought which are connected together in such a way that through this doctrinal model we allow our understanding to come under the articulate revelation and power of God's own Reality and seek to grasp in our thought as much as we may what God communicates to us of his own unity and simplicity. What we are concerned with here is not any conceptual system as such, but the immanent relations in God himself as Father, Son and Holy Spirit in the communion of love and knowledge which he sets up between us, and it is with that end in view that we proceed to develop a doctrinal system of the greatest conceivable unity and the greatest paucity of fundamental concepts, through which we can provide a revisable conceptual representation of the ultimate intelligible basis of our knowledge of God.

In doing this, however, we have the same difficulty with primitive and pre-scientific concepts and the kind of language that goes with them — from which we cannot finally escape — as in its own way physical science has when trying to grasp and represent the inherently imperceptible realities and relations that come to knowledge in relativity and quantum theory. What we must never allow ourselves to do is to forget the constitutive elements in God's self-communication to us and in our experience and apprehension of him which gave rise to those concepts, even if we have difficulty in representing them to ourselves without what Einstein in a different context called "the spectacles of the old-established interpretation".[2] This is the difficulty that constantly crops up in our having to use terms like "father" and "son" in the formulation of a doctrine of the Trinity which, nevertheless, can be valid for us only in so far as it remains correlated with our fundamental experience of God and controlled by his self-revelation to us in Jesus Christ the incarnate Son of the Father. We can no more get away from using the expressions "father" and "son" than we can do without God's self-revelation as Father through the Son and in the Spirit.[3] Yet it is important to note that the concepts and expressions we employ can be of genuine theological significance for us if they direct us away from themselves to the ineffable Reality of God himself who is greater than we can ever conceive, for they fulfil their God-given function in revelation when they enable us to grasp something of the inner relations of God which are the fundamental relations in and behind our experience of him and are independent of our concepts and expressions referring to them. Thus the inadequacy of our formulation of the Trinity of God is an essential element in its truth and precision, that is, in constituting *not a picturing model* with some kind of point to point correspondence between it and God, but *a disclosure model* through which God's self-revelation impresses itself upon us, while discriminating itself from the creaturely representations necessarily employed by the model, and so bears upon our minds that its own inner relations are set up within them as the laws of our faithful understanding of God.[4]

If then the conception of the Triunity belongs to the ultimate structure or grammar of our knowledge of God, it is impossible to keep separate in two different movements of thought or in two conceptual systems, knowledge of the one God and knowledge of the trine God, as advocated by St. Thomas Aquinas, for that would certainly import a fundamental split in our basic concept of God.[5] It is at this point that we can discern the effect of rejecting a natural theology as an antecedent conceptual system independent of actual knowledge of God, and of reconstructing natural theology within positive or revealed theology in much the same way in which geometry now becomes a form of natural science in the heart of physics.[6] The problem that confronts us here, however, is that of assimilating the radically new conceptions in the doctrine of the Triunity of God into our pre-theological and pre-scientific understanding of him, not least, the problem of reconstructing our earlier conception of the unity of God in terms of what Leonard Hodgson has called "internally constitutive unity".[7] What we have to do with here is a profound unity in trinity, unity at a deeper level than is mathematically conceivable but which is forced upon us by the inner intelligible relations of the one God.

An examination of the history of scientific theories reveals how difficult it has been again and again for distinctly novel elements to be brought into coherent relation with prior thought. Even in the case of some of our greatest and now universally accepted theories there has frequently been long resistance. We recall the long time it took for the Faraday–Clerk Maxwell conception of a continuous indivisible field of force to be accepted as an independent reality in its own right, or the lapse of time it took in our own century for the scientific world to recognise Max Planck's discovery in regard to black body radiation which in its failure to satisfy the demands of Newtonian mechanics gave rise to quantum theory. Whenever radically new departures of this kind have been followed up, they have forced through a deeper, richer, reconstructed understanding of the unity and theoretic basis of physics. Appeal may be made in this respect also to another form of the same analogy taken from musical composition, for example, by Beethoven. Again and again quite new patterns are brilliantly injected into the development of a symphony which at first appear radically different and even quite contradictory to what has gone before, but then they are marvellously developed and woven into the composition in such a way that the whole symphony takes on a greater depth and richer fulness of meaning.

It is not otherwise with Christian theology in its deeper and richer understanding of the oneness of God in the light of the Trinity. How then do we come to epistemological terms with the Triunity of God? The trinitarian structure in knowledge of God is not something that can be inferred from what we may claim to know already, nor can it be logically developed out of prior patterns of thought, but new trinitarian elements can be assimilated into our knowledge through a profound reconstruction which gives knowledge of God a depth and fullness correlated with experience which it did not have before.[8] Here our rational understanding takes on the imprint of what it comes to know and so yields to the Trinity of God's self-communication as Father, Son and Holy Spirit, a structure in the mind that faithfully answers to it. This is natural, for the knowledge of God requires for its actualisation an appropriate rational structure in our cognising of him, but that is not, and cannot be, an autonomous rational structure developed independently on the ground of "nature alone" or apart from the active self-disclosure of God. Rather does it arise under the pressure of the intelligible content of God's articulate self-revelation to realise itself in our faith and understanding, and only as we allow our minds to fall under the compulsion of God's being who he really is in the specific acts of his revelation to us in space and time. As such it cannot be derived from any analysis of our autonomous subjectivity.

It is incumbent upon us, then, to inquire into the unavoidable epistemic coordination between what is known and our knowing of it, between the intelligible features inherent in the object known and corresponding modifications in the understanding of the knowing subject, within which an appropriate rational structure emerges or should emerge. This is a coordination which is unavoidable in any science, although in some sciences where we have to do with what is impersonal or even abstract the need to inquire into it may be minimal, but in others such an inquiry in deference to the very nature of the object or field of knowledge must play a far larger part if only in the service of clarity, objectivity and precision in explanatory description. This is certainly the case with theological knowledge, where explicit treatment of the relation between God and the human agent, the objective and subjective poles of knowledge, is of considerable importance. The subjective pole may not be a sufficient condition for knowledge but it is certainly a necessary condition which as such requires to be subjected to critical reflection and clarification. The spot-light of examination must be concentrated upon the rational structure of the understanding that arises in actual knowledge of God so that it may be methodologically "isolated" or thrown into explicit relief in the light of the divine reality which evokes and conditions it and which it is enabled to reflect as its object.[9]

This must not be taken to imply that the rational structure in question can be disengaged from the actual relationship of knowledge between God and ourselves, as if it could have a validity of its own quite irrespective of any objective reference or content, for it could be no more than a merely formal, empty conceptual husk. On the other hand, if we accept that in actual knowledge there is, and must be, a real relation between God as he is known and our knowing of him, this must not be taken to imply that the rational structure that arises in our human knowledge of God has any prescriptive force. Far from holding that God must submit to the necessity of our human conceptions taken up in that rational structure, we must recognise that as the controlling reality who conditions the rational structure in our knowledge of him, God is himself the source of the necessity in the human conceptions employed in it. What we are concerned to do, then, is to reach deeply into the heart of our intellectual experience of God as he actually makes himself known to us and grasp within it both something of the divine Reality which constrains and conditions our knowledge of him and something of the appropriate rational structure necessarily arising in our understanding as it faithfully reflects him as its object.

In order to focus the problem more sharply, let us compare the ways in which St. Augustine and St. Thomas Aquinas treat the knowledge of the Trinity.

According to *St. Augustine*, we know God by an interiorising movement of thought not to subdue understanding of God to the patterns of our subjectivity, but to let the patterns of our understanding take shape under the impact of his reality. We really know God when knowledge of him strikes into the roots of our personal being and affects the structures of our consciousness. Since God is a Trinity of Persons, our basic knowledge of him must be actualised in and through triadic structures in our knowing of him. Hence St. Augustine gave a great deal of thought to correlations between the divine Trinity and various triadic patterns in human subjectivity, in the soul. This was the form which his "natural theology" took, for it was not conceived as arising on some independent ground and as developing an autonomous structure of its own.

For *St. Thomas*, on the other hand, knowledge of the one God could be reached through natural theology alone, operating on the independent ground of reason and sense experience through causal and analogical inference, whereas knowledge of the blessed Trinity could be gained only through the instruction of faith operating beyond human sense on the ground of divine revelation, for only through faith is the human reason adapted to know God in accordance with his nature.[10] Through this double approach the understanding of the Oneness and Essence of God as a whole became detached from real knowledge of him in his saving and revealing activity in history, so that it was turned into an abstract metaphysical construct and made into a prior independently conceived framework of thought with reference to which all other knowledge of God, even that derived from divine revelation, was bound to be interpreted and influenced. The overall effect of this ecumenically was sharply to divide the Latin approach to the doctrine of the Holy Trinity based on an essentialist account of the unity of God from the Patristic and Eastern approach to the three Persons, Father, Son and Holy Spirit, one Godhead, but its effect epistemologically, as Karl Barth has shown us so clearly, was to introduce a deep split in the fundamental concept of God.[11]

Behind this Thomist and Latin account of the doctrine of God there evidently lies a form of the damaging dualism which we have had to reject in favour of a realist and unitary approach to knowledge in all areas of human scientific inquiry. It cannot be otherwise in a faithful knowledge of God which must be governed throughout by the reality of his self-communication and self-revelation to mankind in history as Father, Son and Holy Spirit. Through his saving and regenerating activity this self-disclosure of God strikes through man's darkness and distortion into the innermost centre of his personal being and brings about within it a profound reorientation and modification of his life and thought, so that from the very start his knowledge of God takes on an inner organisation that bears the imprint of God's trinitarian operations. If the whole economy of redemption has a trinitarian structure, the fundamental activity of human faith and understanding evoked by God's saving work must also have a trinitarian structure, the latter answering to and reflecting the former. This would seem to send us back to St. Augustine to have another look at his approach in attempting to clarify the inner link between the Triune God whom we are enabled by divine grace to know and the structure of our human subject-being that arises in the actual course of our knowing him. Further analysis of St. Augustine's teaching would seem to call for both appreciation and criticism.

Appreciation. The basic problem, as St. Augustine saw it, was how to distinguish the terms "Father", "Son", and "Holy Spirit" from their pre-theological use, and therefore how to interpret them without being ensnared in the primitive images associated with those terms in our everyday life and thought. That is to say, he had to carry through a theological refinement of the basic concepts "Father", "Son", and "Spirit" while remaining within the frame of experience of God which demanded their employment. His other problem was how to conceive the unity of God the Father, Son and Holy Spirit, without detracting from the distinctive reality of the three Persons who are mutually and coeternally interrelated in the Trinity, while all the time giving the utmost respect to the transcendent, ineffable mystery of God which cuts off any possibility of an abstract-essentialist or a logico-analytical approach to an understanding of the Holy Trinity. St. Augustine sought to meet his problems (at least in part) with the Nicene principle of "consubstantiality". This allowed him not only to hold that there is a relation of indivisible oneness in being between what God is toward us as Father, Son and Holy Spirit and what he is eternally in himself, and vice versa, but also to hold that the Father *is* the property of being "father", the Son *is* the property of being "son", and the Holy Spirit *is* the property of being "spirit". That is to say, the so-called "attributes" of "father", "son" and "spirit" predicated of God as his properties are fundamentally rethought in *onto-relational* terms and are thereby refined, with the primitive images which those words carry in their mundane non-theological use thought away.[12] Far from "fatherhood", "sonship" or "spiritness" being projected with their creaturely content into God, they have themselves to be understood from a centre in God.[13]

When in this light we turn back to consider St. Augustine's *De Trinitate* again, we find developed in it a sustained movement of thought of great brilliance and profundity in which he tried again and again to find the appropriate conceptual instrument for an understanding of the divine Trinity which would both bring about a refinement of theological concepts and reveal the intelligible basis upon which they rested in God himself.[14] He selected and tried out different triads of thought which he might employ as models for his theological inquiry, and cast them aside one after the other when they proved unhelpful. Eventually he found what he considered to be helpful triads: being, knowing, willing; lover, beloved, love; mind, knowledge, understanding; memory, understanding, will. The most significant of these St. Augustine felt to be the triadic structure of love, for it evidently derived from the fact that God *is* Love and appeared to reflect and point back however slightly to the supreme Communion of Love which God has revealed himself to be as Father, Son and Holy Spirit. With the help of this basic clue he set about organising these triads into disclosure systems through which the trinitarian structure of God himself might be allowed to shine through and be reflected more explicitly in his understanding. And then in the course of the *De Trinitate* we find St. Augustine developing his thought by means of these triads through three successive stages or levels, moving from one level to the other, until he achieved a compound triadic model in terms of which he was able to correlate human understanding of God with his Trinitarian nature, or rather to discern the kind of triadic structure arising in the human soul which knowledge of the Trinity requires in order to be actualised, and in a sense naturalised, in the human understanding.[15] Built up in this way the *De Trinitate* is certainly one of the greatest books of theology ever written. Its influence has certainly been quite incalculable.

Criticism. It must be granted that St. Augustine's real interest was not to detect the so-called *vestigia trinitatis* or traces of the Trinity in the soul, but to find the adequate concepts and terms through which we can think intelligibly of God as he has revealed himself to us. However, St. Augustine did not work so much with the rational structure of faith, that is, with the kind of rational structure that becomes embodied in our understanding in the actual event of our knowledge of God, or, otherwise expressed, the rational structure that arises from the modification of the human reason under the creative impact of the divine object of knowledge. He worked rather with a rational structure that is found already embedded in the constitution of human nature disengaged from what arises in the actual event of knowing God, and that, he held, must be regarded as a necessary precondition for the realisation of knowledge of God in the human mind or soul.

St. Augustine set out his position succinctly in the following passage from the fourteenth book of the *De Trinitate*.[16] "Although the human mind is not of that nature that God is, nevertheless the image of that nature than which none is better is to be sought and found in us precisely where our nature has nothing better. But the mind must be considered first as it is in itself *before it participates in God (antequam sit particeps Dei)* and his image be found in it. For, as we have said,[17] although it is worn out and deformed with the loss of that participation in God, it nevertheless remains the image of God. For it is his image in so far as it is capable of him and can participate in him; nor could there be anything so very good except in virtue of the fact that it is his image. See then, the mind remembers itself, understands itself, loves itself: if we discern this, we discern a trinity, not yet indeed God, but already an image of God." Alongside that passage must be placed another from the same book.[18] "The trinity of the mind, therefore, is not the image of God, because the mind remembers itself, and understands and loves itself, but because it can also remember and understand and love him by whom it was made." There is evidently an ambiguity in St. Augustine's thought, for on the one hand he refers to the image of God in the human mind before it participates in God, and on the other hand refers to it in terms of its objective relation to God.[19] Apparently St. Augustine did not have in mind, then, any outright identification of man's rational nature with the image of God, so much as man's ability to use his rational powers for knowledge of God. Nevertheless, it was in the reflexive property of those powers that he detected the functioning of the image of God. As Leonard Hodgson expressed it: "An image of the divine Trinity is to be found in the human soul when that soul is engaged in the purely intellectual activity of self-knowledge. For that activity involves the knowledge which begets what is known of it and the activity of the will which unites begetter and begotten."[20] This emphasis upon self-consciousness in acts of self-remembering, self-understanding, and self-loving, even when the soul is engaged in remembering, understanding and loving the Creator, shows that St. Augustine was concerned with a largely psychological analysis of the properties and functioning of the human soul.[21] The upshot of this was to give the structure of human subjectivity too determinative a role in knowledge of God, with some serious consequences even in regard to a basic concept like that of the Word of God,[22] although St. Augustine was nevertheless strongly critical of any attempt to project human distinctions or creaturely features into the divine Being. It was in fact due partly to his concern to avoid anything of this kind that St. Thomas was prompted to assign the doctrine of the one God and the doctrine of the Trinity to different conceptual systems.

It is toward the end of his great work that St. Augustine touched upon what seems to me to be a right clue for deeper theological penetration, which he speaks of as "wonderfully ineffable or ineffably wonderful": "While the image of the Trinity is one person, the supreme Trinity itself is three persons (*cum sit una persona haec imago Trinitatis, ipsa vero summa Trinitas tres personae sint*), yet that Trinity of three persons is more indivisible than this of one."[23] We express this the other way round by saying, "The supreme Trinity is three persons, but the image of that is one person". Now as the history of thought shows, it is the *concept of the person* which is the radically new element introduced into the furniture of human thought by the Christian understanding of God as Trinity. That was a point taken up and discussed in different ways in three series of notable Gifford Lectures delivered in Aberdeen. In the first of these, dating from the sessions of 1912 and 1913, A. Seth Pringle-Pattison had this to say: "The essential feature of the Christian conception of the world, in contrast to the Hellenic, may be said to be that it regards the person and the relations of persons to one another as the essence of reality, whereas Greek thought conceived of personality, however spiritual, as a restrictive characteristic of the finite — a transitory product of a life which as a whole is impersonal."[24] This was followed up by W. R. Sorley, in the series immediately following, of 1914 and 1915, who claimed that "Relations are as necessary to the existence of things, as things are to the existence of relations. Both are required in the constitution of the real."[25] He argued that it is in this way that we must regard the relations between persons, and that as such persons in community are the bearers of value.[26] Then in the series of Gifford lectures delivered in 1918 and 1919 under the title *God and Personality*, Clement C. J. Webb reinforced the claim that it was only "in connection with the doctrine of the Trinity that the words 'person' and 'personality' came to be used of the Divine Being",[27] but went to argue, rather against the tradition of Latin theology, that God himself must be regarded as Personal and indeed as Person, for what God is in his personal relations with us and in our experience of him, especially in worship, he must be within his own divine Life.[28] This must not be allowed to detract from the conviction that the Father, the Son and the Spirit are themselves personal and Persons, but to think of God as Person and of three Persons, one God, we need to give full value to the fact that the relations between persons are no less real then the persons themselves. The Christian concept of person implies a plurality of persons.

It is no doubt somewhat difficult for us now to realise the immense change in the orientation and character of human thought that the introduction of the concept of the person brought about. This was a change, as Pringle-Pattison pointed out, from a basically impersonal mode of thinking to a personal mode and from an impersonal to a personal conception of reality. Today we have taken this change for granted, but it was the direct result of an understanding of God found only in Christianity among world religions, and specifically of the Christian doctrines of the Incarnation and the Holy Trinity. However, once the notion of *person* was produced and launched into human thought, it inevitably came to have an independent history of its own, and has now of course become a universally accepted category of thought. This is not to claim that the notion of person is everywhere understood in precisely the same or in a uniform way, far from it, for the ancient impersonal outlook is still deeply entrenched in our thinking.

Nevertheless, the introduction into human thought of the category of the personal has far-reaching consequences, of a general and a particular kind.

1. If personal relations belong to the structure of reality itself, then it is surely the model of the personal agent that must be primary in our attempt to think intelligibly of God, and not the impersonal model of the detached observer over against the object, with its unbridgeable gulf between subject and object, which has returned to exercise so much influence in our Western thought since Descartes. In contrast, the effect of Christianity is to replace the impersonal *Id* with the intensely personal *Ego Sum* of the living God, but that brings the Christian faith into a wide-ranging struggle with the ancient impersonalism that still exercises considerable if inertial force in sensitive areas of our culture and way of life.[29] Our immediate concern here is with the reforming of our basic conception of God by using the concept of the person or rational agent as a disclosure model through which to allow God's own ultimate personal and personalising nature to contol and shape our understanding of his dynamic interaction with the world he has made. As we have already seen, this radically alters the whole slant of "natural theology", and it is bound to affect very deeply other areas of our modern thought and culture, as is evident in the various writings of people like John Macmuray and Michael Polanyi.

2. The concept of the person and of personal relations in God and in ourselves demands of us fuller consideration than we have given it hitherto. If God's personal self-communication strikes into the innermost centre of human being, personalising it, then it is through an apprehension of that modification of personal being that we may be able to think more intelligibly of God who is transcendently in himself what he is in his personal relations with us, and as

such is the one personalising Person who is the creative source of all our personal relations with him and among ourselves. This is the line of thought that St. Augustine did not take, which is understandable in view of the habit he had of interiorising his thought so that it became turned in upon itself thereby damaging its transcendental structure in relation to the triune God. St. Augustine's rather psychological, interiorising approach to the truth, by looking for it within the depths of his own spiritual being,[30] meant that he had to fall back upon some sort of ontologistic participation by the light of the human mind in the Light of God, the theory of illumination which evoked the critical reaction of St. Thomas.[31]

Now quite clearly a great deal depends on what we mean by *person*. In order to clarify our understanding of the concept I would like to discuss two principal definitions of person which have been put forward in the history of thought, by Boethius in the sixth century, and by Richard of St. Victor in the twelfth century.

According to *Boethius*, "person is the individual substance of rational nature" (*persona est rationalis naturae individua substantia*).[32] This is a concept of person that was *logically derived* from the notion of universal substance. Presupposing the distinction between person and nature, he went on to distinguish between two kinds of nature, substance and accident, in the Aristotelian mode, and argued that only substances can be persons. Then, starting from the universal genus of substance, Boethius distinguished between substance and rational substance, and, making use of Aristotle's distinction between primary and secondary substance, he distinguished between general and particular substance, and so reached the conclusion that "person is the individual substance of rational nature". By individual substance he meant that which cannot be further divided, that is, the concrete particular of rational nature, Plato, for example, or Cicero.[33]

Here we are given a definition which starts from one essence, but of which it is held to be a particular determination. The definition of person thus reached, however, is particular and restrictive, the isolated individual, "the atom", so to speak, of rational being or nature. As such it is an impersonal concept. But when one adds to it, as happened in the history of thought, the notion of an individual centre of consciousness, and of self-consciousness, not to speak of self-expression and self-fulfilment and so on, then one can see how many problems it produces, for the person is defined here in terms of itself in its cut-off particularity and private individuality.

An approach to the doctrine of the Holy Trinity on these lines could only have the effect, and did actually have the effect, of throwing into a hard logical form, the typically Western and Latin doctrine which started from one divine essence or substance, and so distanced it even further from the Greek patristic approach which started from God the Father and the coeternal and consubstantial communion of Persons as the ineffable Trinity.[34] Apart from that, however, the restrictive concept of person thus reached is very difficult to apply to the Father, the Son and the Holy Spirit in their wholly interpenetrating or "perichoretic" relations, as the three Persons of the Holy Trinity were understood in the high Patristic theology from Athanasius to Cyril of Alexandria that lay behind the Ecumenical Councils. Nor of course can the Boethian concept of person be applied to God as such, who is not an individual person. But neither can it properly be applied to human subject-beings, for it shuts the individual up in himself, so that his natural movement is one of self-determination over against other isolated individual subject-beings. However, difficult as it is, this is the notion of person which has tended as a matter of fact to dominate the main history of thought down to our own day, and has led so many, Hegel or Tillich, for example, to deny that God can be spoken of as personal. If this concept is applied to God it would seem to mean either that God is a restricted individual or that there are three finite Gods.

According to *Richard of St. Victor*, on the other hand, "person is the incommunicable exsistence of intellectual nature" (*persona est intellectualis naturae incommunicabilis exsistentia*).³⁵ This is a concept of person that was *ontologically derived* from the Holy Trinity, in deliberate rejection of the Boethian concept. Richard did not try to reach this concept through logical processes, but through allowing the unique reality of personal being to stand out (*ex-sistere*) so that it could be apprehended in its ontological relations out of its own proper ground in the Trinity. And then in that light he essayed a definition of person which could apply appropriately to the three Persons of the Trinity on the one hand, and in all due difference appropriately to human persons on the other hand. He defined the person, therefore, not in terms of its own independence as self-subsistence but in terms of its ontic relations to other persons, i.e. by a transcendental relation to what is other than it and in terms of its own unique incommunicable exsistence, i.e. in terms of what Greek theology spoke of as *hypostasis* which had been literally but misleadingly rendered in Latin by *substantia*. A person can communicate with others but it, or rather, he retains the inalienable mystery of who he is in his own distinctive reality which may not be resolved away or be overwhelmed by the subject-being of the other. Nevertheless a person is what he is only through relations with other persons. Thus the incommunicable exsistence represents the fact that a person is really objective to what is other than he and that this objectivity of one person to another is a constitutive ingredient in personal being: *proprium non est ex se sed aliunde*.

Richard distinguished, however, between *communis exsistentia* and *incommunicabilis exsistentia, communis exsistentia* denoting in the first instance the Holy Trinity, and *incommunicabilis exsistentia* referring to the Persons of the Trinity. *For God* this means that as the Trinity he is a fullness of Love and of personal Being, and as such is the creative, archetypal Source of all other personal beings and their interpersonal relations of love. Thus all other personal being must be defined in terms of its whence and whither, its source and its end, that is, by reference to the fulness of Love and personal Being in the Trinity. *For man* the distinction between *communis exsistentia* and *incommunicabilis exsistentia* means that his essential personal being is not to be seen in its individuality, its self-subsistence or its self-belonging, but rather in the whence and whither of the self, in the Communion of Love and Personal Being in the Trinity. This means that man's essential personal being is defined, through relation to God, by what is not-self, by reference to its objective source and its objective end, for example, in loving the other objectively for the other's sake. It is precisely in this kind of out-going ontological relation and not in his self-contained being, that is, in the onto-relations of love, in person-respecting and person-evoking relations, that a person has his own inviolable dignity, freedom and reality.

In other words, the concept of person is defined here in such a way as to make clear that a person is what he is by way of reference to God and his neighbour at the same time. It is within those very relations that he "exsists" or stands out as what he is in his unique reality and freedom as a person. As such, however, personal being is essentially *open* to others and is found only in intercommunication and incommunicability: the personal and the social belonging essentially and inseparably together. Understood in this way it becomes clear why the concept of the personal arose in human thought out of Christology, the doctrine of the acute personalisation of God's relations with human beings in Jesus Christ, and out of the Trinity, the doctrine of God as the eternal Communion of Love which is the creative ground of all true personal being among men and women, and thus out of the teaching of the New Testament that we cannot love God with all our heart and with all our soul and with all our mind, without also loving our neighbours as ourselves.

Strange as it may seem, it is not this Ricardine but the Boethian concept of the person that has tended to dominate the history of thought, although through St. Thomas Aquinas, doubtless under the influence of Alexander of Hales, his Franciscan teacher, the Boethian definition attracted to itself the notion of "incommunicability" from Richard of St. Victor. Nevertheless, I believe that this concept of the person which, as Richard showed, derives from the Triunity of God, gives us a more helpful lead into an intelligible grasp of the relations between us and the Triune God. However, it requires to be sustained by a fuller doctrine of the Holy Trinity in which the emphasis of Greek theology upon three Persons, one Godhead, is brought in to redress the emphasis of Latin theology on one God in three Persons, in such a way that our understanding of the personal Being of the one God and our understanding of the Communion of Persons in Holy Trinity reinforce and complement one another.[36] This would have the wider effect of putting even more pressure upon us to abandon any impersonal model of thinking of the relation of God to the world, together with the objectivist observer approach and the deistic dualism that have traditionally gone along with it.

Now let us return to the point that if our rational understanding takes on the imprint of what we know, knowledge of the Holy Trinity must yield a structure in our mind or soul that answers to the Trinity. In exploring that we took our initial clue from St. Augustine's idea that while the supreme Trinity is three Persons, the image of the Trinity is one person. Our discussion of St. Augustine's thought and of that of Richard of St. Victor particularly has brought home to us the realisation that what images the Trinity is our interpersonal structure, and not least the interrelations of love which reflect the fact that God is Love in the consubstantial Communion of Father, Son and Holy Spirit, although in the nature of the case our inter-personal relations of love have properly to be understood from the Communion of Love in God which is both their source and their goal. In the light of that discussion we may now go on to probe further into the relation between God as he makes himself known to us and our human knowing of him, and between God known as the personalising Agent and human knowers as those who are personalised through communion with God and who live in person-generating relations with one another. Let us pursue this in several steps.

1. Through personal interaction with us God creates reciprocity between us. By encountering us as personal Being God at once brings us into a personal relation with himself and prevents us from including him within our own subjectivity, for it is as the Thou, the transcendent Other, that he meets us and makes himself known.[37] He both distinguishes himself from us as independent Reality over against us, and indeed as Lord God of our very being, and at the same time posits and upholds us before him as persons in relations of mutuality and freedom with God and with one another.

There are several elements here which I would like to stress with some reference to the thought of Martin Buber who as a Jewish thinker was not of course a trinitarian.[38] First, we find ourselves caught up in a relation of compelling obligation which touches us at the very root of our being. We have to do with the downright Reality of God who lays absolute claims upon us which we have to acknowledge and affirm absolutely. This is not the kind of relation over which we can exercise some kind of control, as though we were free to assent or not to assent it, for what we do can only be done out of a binding relation to the Truth itself. Secondly, we find ourselves prized open, as it were, for face to face with God our monologue with ourselves is turned into dialogue with another. The circle of our subjectivity is breached. Our self-centredness, our selfishness, our introverted self-affirming existence are under attack, and we are taken captive in a transcendental relation to what is beyond us. But in this very movement of extroversion we find ourselves being healed of a damaged personal structure. Through the re-establishment of objective relations to an independently other, the personal self stands out more and more in its unique and inviolable dignity and spontaneity. Thirdly, within the reciprocity set up between God and us in this way there arises a compulsive anthropomorphism, for it is in turning himself toward us that God turns us to himself, and in adapting his own ways to us that he adapts our ways to him. As Martin Buber reminds us, "it is in the encounter itself that we are confronted with something compellingly anthropomorphic, something demanding reciprocity, a primary Thou".[39] This is an anthropomorphic element that remains in our basic concepts for it arises within the actual frame of experience in which we meet God even in his absolute independence of us, and expresses the concrete quality of the astounding reciprocity evidenced and sustained in the encounter. That God comes to us and meets us in the midst of the natural objectivities and intelligibilities in which we are already locked, even though in opening them out and adapting them for the purpose of making himself known to us, is not without its difficulty, for we have to learn how to cope with the primitive and preconceptual images that cling to this experience and must be constantly aware of the temptation to project their creaturely content into God. On the other hand, it should also become clear to us that we cannot shed the anthropomorphic features entirely without retracting out of the intimate personal reciprocities at the root of the very knowledge of God which he mediates to us. Moreover, to retreat from that encounter with God into some encounter with ourselves is to abandon ourselves to the destruction of our essential personal structure, for it is to cut ourselves off from person-generating relations with God.[40] That is why, no doubt, modern reaction against the Incarnation as the centre and core of God's ultimate self-communication to mankind, whereby he has intensely personalised his relations with us in space and time, so bankrupts the personal and social structure of human relations with God by emptying them of their constitutive relation with him, that the miserable cry can go out that "God is dead". To reject reciprocity with God at this concrete, decisive and archetypal point, cannot but have a retroactive effect in undermining the very ground on which inter-personal relations may find emancipation and regeneration.

To return for a moment to the "compelling anthropomorphism" of which Buber spoke, we must not forget that the downright Reality of God within encounter means that by his transcendent objectivity God sets limits to the reciprocity he establishes between us. Thereby he sets boundaries to our knowledge of him, which prevent us from arguing speculatively or prescriptively out of our own subjective structures and compulsive unavoidable anthropomorphisms. This is why we have to take very seriously the problems which everyday rigorous science has to face as it penetrates through level after level of language and concept into the ontological structures of reality, as far as that is possible. These are the problems that arise from the fact that our language is compelled to work with words inseparably connected with primitive concepts, and with old-established conceptual interpretations carried by the very words we use. The need to purify these concepts, while continuing to use this language, gives rise to the stratified structure of scientific inquiry in which we develop organised systems or patterns of concepts of a refined order which are nevertheless correlated with the basic frames of experience in which our basic concepts arose. That is precisely what we are concerned with in the doctrine of the Trinity, in which our concepts of God as Father, Son and Holy Spirit are refined from the creaturely images latent in our ordinary use of those terms, and yet are developed in such a way as to deepen our hold upon the original experience that gave rise to them. In this way the *doctrine* of the Triunity of God both serves as a disclosure model through which the objective depth of divine Reality becomes more and more disclosed to us, and at the same time advances the empirical development of our personal and social structures in a communion of love. Thus while the doctrine of the Trinity puts limits to our human understanding of God, it also generates the notion of person that guards the mystery and inviolable dignity of the other as something which we must respect for its own sake. But this is to anticipate the next step in our discussion.

2. Within the reciprocity of which we have been thinking God communicates himself to us in a movement of love which penetrates into the structures of our human

existence and sets up within them the law of love in its own internal relations. It creates a circle of knowing which is also a circle of loving, in both cases a circle resting on the free ground of God's own trinitarian Being and Activity. But this circle of knowing and loving gives rise to a community of reciprocity in knowing and loving. That is to say, God's self-giving and self-revealing not only penetrate into the innermost centre of the human person but into his fundamental relationships in the world of space and time, and sets up within them a community of personal relations, in which the objectivity of each person over against the other is respected in love and in which they become open to each other in inter-personal and social structures which reflect the Communion of Love within the Holy Trinity himself. We may express this the other way round, by saying that the Communion of Love in God has inter-penetrated our human existence in such a way as to generate within it a community of love which participates in and is sustained by God's own Communion of Love in the consubstantial and interpenetrating relations of Father, Son and Holy Spirit. It is within this community of reciprocity with the Trinity that we come to realise not only that God himself eternally lives in and is a Communion of mutual Love in himself, but that the very core of personal being which we derive from him involves a movement of mutual love among us in the out-going and responding of persons to one another. In the nature of the case this excludes any notion of the person as an isolated individual whose essential movement is grounded on himself in the form of self-love, self-encounter, self-fulfilment, and so on. That is to say, the personal God who communicates himself to us in revelation and love, and a community of love among human beings, belong insepar-ably together in our knowledge of the Triune God, for it is precisely in that community where the Love of God has set up its own inner relations as its determining structure that we may grow and develop in our knowledge of him as a transcendent Fullness of Love in himself and as the constituting ground of all authentic relations of love among human beings. It is within such a communion of love, then, that personalising, person-constituting relations heal our damaged personal and social structures.[41]

Now as we reflect upon this experience of God in which his self-giving and self-revealing draw us into a circle of knowing and loving which takes in God and ourselves, we find ourselves enveloped by a knowledge of God which opens out into the eternal relations within the divine Being. We are brought to know God in such a way that we know him to be in his own eternal Being the Love which God the Father is toward us through the Son and in the Spirit, and know his inner trinitarian Being to be an eternal movement of Love. The knowledge and love of God that arise within this communion of God and interpenetrate our personal relations in a reciprocity of love, bear in them the irreducible conviction that God is in himself a plenitude of eternal relations of Love.[42] The Love with which we are loved by God is the Love with which the Father, Son and Holy Spirit love one another in the Trinity, and the knowledge with which we know him is a sharing in the eternal knowing in which the Father, the Son and the Holy Spirit know one another coeternally. What God is toward us in the structure of love relations which he has set up between us and himself, he is in himself as God. This is only to spell out a little the cardinal issue brought forward by the Church Fathers in the fourth century as they sought to distil out of their interpretation of the Holy Scriptures the essence of the evangelical message that God so loved the world that he gave his only begotten Son for our salvation, namely, that God is antecedently and eternally in himself the Communion of Love which he has manifested to us in his revealing and saving acts.

3. So far in our discussion we have paid little direct attention to the Incarnation itself in which the cardinal issue we have just mentioned took concrete form. Here we have to do with the self-giving of God to us within the objective and contingent structures of our worldly and historical existence, the movement in which God himself in his self-giving entered into the heart of our personal and

social being in the form of embodied Love without any diminishment of his divine Reality. By communicating his own Self to us within our personal and physical existence in this way God has, as it were, externalised himself for us within space and time, and thus consecrated the ground where he continues to meet us directly and personally within our human being in Jesus Christ, and where through his Holy Spirit poured out upon us God lifts us up to share in his own divine Life and Love. It is in Jesus Christ that the Word of God became flesh in human history and penetrated our human forms of thought and speech, but in such a way as to interpret God to man and man to God, bring God near to man and bring man near to God, mediating the one Spirit of God to man and adapting human nature to receive and be indwelt by that Spirit. Since it is through Jesus Christ and in one Spirit that we are given access to God the Father, it was Christology that became the natural starting for a doctrine of the Trinity.

The trinitarian understanding of God was already implicit in the mind and worship of the Church from the beginning, as we can see in various New Testament formulae which bring Father, Son and Holy Spirit together in the Name of one God, not least that used in the sacrament of Baptism.[43] But to discern how this understanding took explicit form in Christian theology, it may help to consider some of the basic questions forced upon Christians as they reflected upon the fact that, centering in the life, death and resurrection of Jesus Christ, God had made himself known in saving and sanctifying acts as Father, Son and Holy Spirit. These were questions about the actual content of what is revealed and communicated, about the agents of this revelation and communication, and about the intensely personal and spiritual way in which all this took place. And there were also questions about the nature of the relation between God himself and the content and agency of his self-revelation and self-communication, and more specifically about the three modes in which God has made himself known to us and what he is inherently in his own eternal Being as God. In finding that they had to acknowledge that God *is* in himself what he is in his revelation and communication to us, that is, that there is a *oneness in being* between the Being of Jesus Christ and the Being of the Holy Spirit and the Being of God the Father, and yet in having to distinguish between God's self-giving through Jesus Christ and his self-giving in the Holy Spirit, they found that they had to recognise that there are ineffable differentiations within the one Being of God, but differentiations in which the incarnate Son and the Holy Spirit are both God in the same sense in which God the Father is God. Ultimately inexpressible though this Holy Trinity was for them, they were convinced that they had to affirm that the modes of God's self-revelation and self-communication were not just modes, aspects, faces, names, or relations in God's manifestation of himself to us within the transient forms and conditions of our creaturely existence, but were distinct while completely interpenetrating *"hypostatic"* or *personal modes of being*,[44] for both in the incarnate Son and in the Holy Spirit it is not just something of himself that God has revealed and communicated to us but *himself* in his own divine Reality.

Undoubtedly the pivotal point in this trinitarian understanding of God was the Incarnation in which God objectified himself for us in Jesus Christ in such a way that, God though he was, he came among us *as man*, and yet in such a way that he did not, as it were, resolve himself into man without remainder. He came as God and man in the indivisible unity of the incarnate Person of the Son. However, since he came to us in the undiminished reality of a particular human being, that human being in virtue of his oneness in being with God reveals God not just by confronting us with himself but by pointing us away from himself to the Father. Thus while in the Incarnation the self-revelation and self-communication of God took concrete objective form among us in Jesus Christ, that self-revelation and self-communication do not simply reduce to the objective form taken by the Incarnation. Otherwise they would reduce to merely creaturely activities somehow mediating between God and man. Rather in Jesus Christ the self-revelation and self-communication of God are incarnated in such an objectified way that God remains *Subject* and transcendent *Lord* who retains his own incomprehensible mystery and sovereignty. This incarnate self-revelation and self-communication of God are not understandable apart from the self-giving of God to us in the Holy Spirit, for it is in and through the Holy Spirit that God imparts himself to us in Christ in such a way as to lift us up to share in the Communion of Life and Love which God is in his own eternal Being. It is as Jesus Christ and the Holy Spirit mutually mediate one another to us that our knowing of God is not confined to the objectified form of his self-revelation and self-communication but in and through it is made to terminate upon the transcendent Reality of God the Father. Thus the movement of God's revelation and communication of himself to us which completes itself in an answering movement from us to God, takes an essentially triune form: from the Father through the Son and in the Spirit, and, correspondingly, through the Son in the Spirit to the Father. And in that whole movement God discloses to us that he is Father, Son and Holy Spirit, a Holy Trinity, in his own eternal Being, and not just in his economic manifestation to us in space and time.

Earlier, in seeking to apprehend something of the trinitarian structure that arises in a faithful understanding of God, we were led to consider God's self-giving to us in terms of the penetration of his inner divine Communion into our personal and inter-personal structures. We must now try to relate that line of thought to the classical doctrine of the Trinity which we have just been discussing.

Through God's self-communication to us within the personal structures of our human being we are drawn into the "vertical" relation of the incarnate Son on earth to the heavenly Father, and thereby share in the relation of mutual knowing and loving between the Father and the Son. At the same time God communicates himself to us in another act by pouring out upon us the Spirit of the Father and of the Son in such a way that there is set up on the "horizontal" level within our social or inter-personal existence a communion of love as a created counterpart or reflection of the trinitarian Communion of Love within the Life of God. That is to say, through the incarnate life and work of Jesus Christ God reconciles and binds us to himself, and dwells within us through the presence of his Spirit, in such a way that he upholds and sustains our relationship with himself from below within the frailty of our creaturely conditions, enabling us to receive his self-giving in Love. Through the Communion of the Holy Spirit we are given to share in a meeting of God with himself within the structured relations of our personal and social being and are thereby enfolded within the divine Self-Communion of the Holy Trinity. Thus we may say that God meets us and reveals himself to us not only, as it were, as Object-Being, but as Subject-Being, whom we know in being known by him, whom we love in being loved by him, and whom we know and love by being admitted into God's own interior Communion with himself. While each of us is adopted by the grace of God in Jesus Christ to share in the relation of the incarnate Son to the Father in one Spirit, this takes place within the corporate communion of love in the Church which as the Body of Christ both reflects the heavenly Communion of Love within the Trinity and represents an externalisation of that Communion within the personal and social conditions of our historical human existence.

Our concern in this discussion is not primarily with the Christian doctrine of the Trinity itself, but with the reorientation of our outlook and the reconstruction of our ways of thinking that follow from it in the emergence of the concept of person and in the personalising of our life in the world. It may help us at this point, then, to consider the articulate actualisation of God's self-communication within the relations which we human subject-beings have with one another in our common relation to the object-world around us, that is, within the intelligible structures of our subject-subject and subject-object relations, and therefore within the triangular structure in which subject is related to subject through their relations to a common object, which we may speak of as a subject-object-subject structure.[45]

The relations which we have with one another as subject-beings or persons are not unmediated, for they take place through the invariant medium of the physical, determinate world of space and time in which we all exist, and which is necessary as a frame of reference for the public language in which we communicate with one another. Yet the relation of person to person, if not unmediated, is nevertheless an immediate one, as mind meets mind in and through the medium of word and world. As rational beings we encounter one another and know one another directly, and do not just infer one another's existence from the alleged "sense data" of some intermedium, but as embodied minds we have the advantage of sharing together in a spatio-temporal continuum through which our communication with one another is subject to objective control. Thus there emerges a triangular structure of communication in our intelligible relations with one another in and through our common world. It is into this human and worldly situation, and into the conditions and constraints to which it is subject, that God's self-communication has penetrated in the Incarnation of his Son in Jesus Christ. And it is within it that he has established relations of reciprocity with us, and, as we have seen, a community of reciprocity, but this is a state of affairs in which God has to reconstruct not only our relations with himself but our relations with one another and with the world in which we live, by reconstituting their reference to himself. Thus God opens out our relations with him and our relations with one another, and indeed with the world in which he has placed us. It is this transcendent reference that is all-important for our understanding of one another and of the world. The divine self-communication transcends the self-communication of the creature and lifts it above and beyond itself, making it open to others, and open to the world, in what we may call the open-space of transcendence. Personal human beings need space between themselves, public and social space, in order to be free, to be out-going instead of in-turned, else they suffocate and perish.

In this understanding of the personal triadic structure interpenetrated by divine self-communication we avoid personalism on the one hand and behaviourism on the other hand, or existentialism on the one hand and objectivism on the other hand. The objectivity of person over against person is strengthened, and the inviolable otherness of persons in their self-body relations is maintained. Moreover, the reality of the world of nature is respected, for it belongs to the general frame of our personal relations with one another and with God. Thus our binding relation with God will not permit us to misuse the world he has made. The very fact that God reveals himself to us, and gives himself to us, within the space-time structures of our world, where we also communicate with one another, implies that we do not have to with God except as he relates himself to the world, and we do not have to do with the world except as the realm where we have to do with God as well as with one another and where therefore it is invested with a sanction beyond the beauty and harmony of its inherent rational order. These interstructural relations must be regarded as belonging to the rich complex of God's creation which in creating he affirmed as good, but which he has now consecrated and confirmed in and through the Incarnation which tells us that God has not held himself aloof from us in our creaturely, physical and contingent existence, but on the contrary has embodied his Love within them in the concrete personal form of Jesus Christ in whom God has irreversibly pledged himself in support of those structures.

It must be noted that the primacy and centrality of the Incarnation in all our relations with God implies that the anthropomorphic elements inevitably involved in the reciprocity established by God between us and himself, far from being eliminated are actually reinforced, but now it is Jesus Christ, the one image-and-reality of the invisible God who constitutes the critical point of reference to which we must constantly appeal in putting to the test and thinking away all inappropriate and unworthy ingredients in our creaturely and anthropomorphic images, concepts and terms. This applies also to the self-communication of God to us within the space-time structures of our world where spatial and temporal ingredients are inevitably involved in our knowledge of God. As themselves the bearers of our creaturely rationality the space-time structures of the world constitute the rational continuum in and through which we communicate with one another and in and through which God communicates with us. Hence we cannot opt out of them without opting out of the spatial and temporal ingredients in our knowledge of God, any more than we can opt out of the very rational structures in which we have our being, think, live and speak. Thus we cannot but think of the Triunity of God from out of this triadic subject-object-subject structure in our worldly inter-personal existence, but it is one that serves God's trinitarian self-revelation to us only as it is taken in command and adapted by that revelation and thus made to point infinitely beyond itself. Here too, however, it is through critical reference to Jesus Christ, the incarnate Word and Truth of God, through whom God reveals himself to us as the Lord of our being and understanding, that we have to put all spatial and temporal ingredients to the test, thereby filtering away from our conceiving of God all that is inappropriate to or unworthy of the Holy Trinity. Hence, once again, we discern the crucial significance of the fact that in and through the Incarnation God's self-communication and self-revelation to us have penetrated into and taken commanding form within the subject-object as well as the subject-subject relations of our human existence.

In view of this it is hardly surprising that, from early Christian times to our own day, subject, object, and relation, should have been developed as a significant intelligible structure in promoting an articulate understanding of the Holy Trinity. Used merely in this form, however, a subject-object-relation model can be and actually has been severely misleading, for it suggests an understanding of the Holy Spirit only in terms of the mutual relation of Love between the Father and the Son which fails to give full reality to the distinctive, incommunicable personal Being of the Holy Spirit. To a certain extent this may be corrected, as in Augustinian theology, by thinking of the Holy Spirit as "a kind of consubstantial communion of the Father and the Son",[46] that is, in an onto-relational understanding of the Holy Spirit who *is* in his own Person or hypostatic Reality of one and the same Being with the Father and with the Son within the indivisible Communion of Love in the Holy Trinity. But this still seems to fall short of the biblical approach of Greek Patristic theology in which the Holy Spirit as a distinct and fully hypostatic or personal Mode of God's *self*-giving, parallel with but interpenetrating that of the Son, and as God of God in the same way in which the Son is God of God. Thus the differentiation between the two hypostatic Modes of God's self-giving in the Son and in the Spirit in which he communicates *himself* reveals that they are equally and coeternally Persons in his own trinitarian Being. This carries with it, of course, a doctrine of the complete coinherence or interpenetration of the three Persons, Father, Son and Holy Spirit, as one God.

On the other hand, when subject, object and relation are reorganised in the form of subject, object, and *subject*, to be used as a theological structure, interpreted and appreciated in the light of the actual structure of God's two-fold self-communication through the Son and the Spirit, and governed in our understanding by the articulate content of what God has revealed through the incarnate Son or Word,[47] it would appear to provide us with more than a merely disposable model. Then it can be used as a disclosure model through which the hypostatic trinitarian relations of Father, Son and Holy Spirit may be allowed to bear upon our understanding and knowing of the Trinity in a deeply informing and creative way, and is therefore constantly subject to revision in the light of what is disclosed. Since the truth of the model does not lie in the model itself but in the Reality of the Holy Trinity to which it refers, recognition of the inherent inadequacy of the model belongs to the truthfulness of that reference. In handling the model in this way we prevent ourselves from thinking about the Trinity by way of projecting some sort of this-worldly trinity, organised out of our epistemological structures, into God, but use the model to serve knowledge of the Trinity as the ineffable Communion of Persons in God revealed to us as God makes himself the Object of our knowing and loving through Jesus Christ in one Spirit without ceasing to be Subject or Lord.[48] Thus we may think of God as communicating himself to us within the structures of our personal and epistemic relations with one another in this world in such a way that, while in the Incarnation he adapts himself to the weakness of our creaturely structures of knowledge, in the mission and presence of the Spirit he uses them as the very means to lift us up above and beyond ourselves to apprehend and love him in his Triune Being. While God makes himself known to us in this way as he who infinitely transcends all our creaturely conceptions of him, at the same time he also gives us an inviolable pledge that he is not different in himself from what he is in his incarnate Reality in Jesus Christ within the concrete actualities of our human existence.

Let us now reflect a little on the way in which Christian understanding of God as the Holy Trinity bears upon our human culture, in science and philosophy and in our social developments.

Our existence as persons requires an environment characterised by certain fixities, i.e., a "firmament" as the Bible calls it, which can become the general framework for a steady, reliable life and also for universal conceptions and public language through which we may communicate intelligibly with one another, clarify our common thought, order our behaviour in universally acceptable ways, and explore the universe together in community endeavours. We need a world with determinate conditions and limitations in which to live as rational persons — and that is after all what we are, embodied rational agents who have roots and connections in the physical world from which we are not to be severed. However, the fact that the human person needs a physical world of constant and regular features to provide the stage for the fulfilment of his life, carries with it the fact that he is also subject to the constraints of rigid patterns of law and exposed to the tyranny of necessitarian structures. Paradoxically, the more he concerns himself with the understanding and determination of the kind of order immanent throughout nature, the more he seems to become a prisoner of it and not least of the machinations he develops under its control. Somehow the technology he devises assumes a momentum of its own, which we feel most acutely in our modern technological society, for, as Bonhoeffer once expressed it, "Technology is the power with which the earth grips man and subdues him."[49] When that happens, rational human being feels his freedom threatened, his personal structure being undermined, for he is being reduced to being a thing.

As person, however, man is the being who is open to what is beyond himself, open to others as well as to the world. It is indeed through this openness of being to the world that he can easily become enslaved to the determinate patterns of nature. But what he needs in this state of affairs is an openness that transcends the power of the determinate world to fix him in its rigid structures and suppress his freedom. He requires a transcendent reference, an Archimedean point beyond him and beyond the power of his own sophisticated technology to control or master. As we have already noted, a transcendent relation of this kind is essential to the integrity, freedom and continuance of pure science, in dedicated openness to truth over which it has no control. But such a transcendent relation or openness to what is beyond it is also necessary for the existence of authentic personal being. Now it is precisely this kind of openness that is created through the intensely personalising interaction of the Triune God.

Let us approach this from a rather different angle, from the relation of person to word and the development of inter-personal life in language-permeated communities, recalling our earlier discussion of the social coefficient of knowledge. It is through word that human beings are able to achieve a measure of transcendence over nature for it is by means of language that they can represent things to themselves, draw apart from and think about them and decide what they are to do in relation to them, that is, with a space and freedom in which they stand out or "exist" as persons. It is through language, or rather through thought and language operating together, that human beings gain a position of independence over the object-world of their knowledge and the environment of their activities. Thus the culture of language, in spoken or written form, is essential to the development of personal and social being. But word-rationality which has its expression in language is impossible without number-rationality, that is, without the kind of rationality that characterises the determinate structure in which we live and think and behave within the space-time universe. Number is the form of rationality or the inherent order which the material universe displays in so far as it is quantifiable, subject to mathematical analysis and computation. The more we immerse ourselves in physical science the more we are inevitably engaged with number. But number-rationality has a very significant part to play throughout the whole range of our human life and culture, for since it expresses the rational order of the physical framework within which all our thinking, speaking and behaving take place, it exercises a salutary constraint upon our subjective fantasies and provides a controlling foil for the maintenance of objectivity in our thought and communication. However, it is only too possible for the relation between number-rationality and word-rationality to become unbalanced in such a way that the power of word is seriously diminished through the encroachment and domination of number. When that happens human capacity for transcendence and openness is severely lamed and the inner structure of personal being becomes eroded — the person is threatened with being reduced to a thing if not just to a number.

Here we touch upon the crisis of our times, evident not only in the deep splits in our culture, in the disintegration of form in the arts, but also in an irrational, romantic "backlash" against scientific culture, but behind these various states of affairs we find word threatened by number. Think for a moment of the great development of mathematical logic, immensely valuable and quite necessary as it is to philosophy as well as science. It can so bewitch their minds that some of the most eminent of our contemporaries have imagined that in the last resort all human thought-connections could be reduced to a complete and consistent system of logico-mathematical symbolism. Is it not a similar form of this imbalance between word and number that we find in the passion for so-called machine-intelligence? But let us take another glance at Marxism again, the relentless development of the concept of the technological society in which personal and social being are subjected to compulsive determinisms from below. I refer to the Marxist society for another reason, to point to the fact that in those countries where it has been longest entrenched we find a powerful resistance being mounted from the ranks of the *litterati* — one need mention only Solzhenitsyn — that is, those whose personal existence is vocationally bound up with word, and who seek through the cultivation of great literature, without subordinating the message to the medium, to create new centres of freedom, dignity and humanity which may break open the tyranny of "scientific materialism". It is basically a similar enterprise that Noam Chomsky pursues in the field of linguistics, in his struggle with the behaviourists, and perhaps also with himself, and his rather Cartesian concept of innate structures in the underlying foundations of grammar to which he looks for an explanation not only for the universality but for the remarkable spontaneity of language.

If we have Chomsky in mind, should we not ask whether it is possible to solve the problem of how we recognise novelities and create quite new patterns in language by seeking an answer from the depths of our existence?

Certainly the Christian would question the possibility of finding a solution through any deeper interiorising of personal structures within the self in its interrelation with this-worldly reality, and it would seem that our consideration of the rise and development of the concept of person bears that out. It is at this point that the trinitarian structure of thought seems to be so relevant once again, not least in what it imports of the essential transcendent relation of the human person beyond himself. Our personal structures in their integrity, openness and freedom are bound up with a constitutive relation to the unlimited Being and Freedom of God, which makes them open also to the world of other structures in things and persons. When we lose that kind of openness which comes from openness to the living personal God, we lapse back into a flat meaningless world and become quickly incarcerated in ourselves with all the dreary boredom that results from that. This is only too evident in contemporary society, not least among our young people, after the incessant indoctrination to which they have been subjected for so long about the finding of personal meaning through self-fulfilment and self-expression.

I submit that it is only through a divine Trinity who admits us to communion with himself in his own transcendence that we can be consistently and persistently personal, with the kind of freedom, openness and transcendent reference which we need both to develop our own personal and social culture and our scientific exploration of the universe. I believe that it is in a radical renewing of our personal and inter-personal structures that comes from communion with God, that we are to look for a healing of the deep splits which have opened up in our modern civilisation. But this means that what we need is the recovery of *spiritual being*, being that is open to personal reality and not imprisoned in its own self-centredness. Or, if it is preferred, what we need to recover is the out-going movement of subject-being, enabling it to behave objectively toward others, and to engage in objective knowledge of things in the realm of nature without being subject to the tyranny of a mechanistic and deterministic conception of the universe. How is that possible without a transcendent reference in the structure of personal and inter-personal being, and that means, from a theological perspective, without the kind of transcendent reference that comes from the renewing power of the creative Word and Spirit and through them from communion with the Holy Trinity? Human society cannot be transmuted into an authentic community of personal being merely through a redisposition of its diseased in-turned structures, for that cannot offset the steady disintegration and fragmentation that result from the conflict of group-egoisms so evident in our modern world. It is the conviction of the Christian faith that such a transmutation can take place only through the reconciliation of people with God and with one another and through a healing of personal and inter-personal structures in their ontological depths through participation in the creative source and fullness of personal being in the Communion of the Holy Trinity. Human beings need to be turned inside out in a profound inversion of their self-centredness and to be anchored in a transcendent centre of Love in God if they are to be persons freely open to one another and the universe which God has created.

In concluding our discussion, let us relate the point which we have now reached with that at which we started. Is meaning something that we create for ourselves, something that is only the projection of our own arbitrary desires or vain attempts to overcome our transience in this world, or is it something that is disclosed to us within the universe, from beyond ourselves, deriving from the objective structures of reality and giving constancy and validity to the enterprise of human culture and inquiry? We found that it is by adopting the first attitude of mind that human endeavour runs out into the chasm of meaninglessness and futility, but that in taking up the second or classical attitude of mind, by operating under the constraint of the intrinsic intelligibility of the universe which reaches out inexhaustibly beyond all formalised science, art and religion into God, that the great advances in human culture and civilisation have been made. By its very nature the intrinsic intelligibility of the universe is open to a transcendent ground as its sufficient reason and sustaining source. And so I have tried to show in various ways that the rationality with which we are all bound up as personal human beings within the vast and wonderful harmony of the universe cannot stop short in its inquiry at the level of rationality immanent in the contingent processes of the creation, but must inquire into the nature of this state of affairs in the universe which by being what it is in its intrinsic intelligibility is open to explanation beyond itself. And I have also tried to show that the rationality intrinsic to us as personal human beings, in the correlation of our knowledge with the symmetries and invariances of the created universe as they become disclosed to our scientific inquiries, is what it is through constitutive relation to the personal Agency of the Creator who is himself the ultimate Ground of all rationality. It is through the work of scientific theology in clarifying the inner correlation between our human knowing of God and God himself, and thus reaching more and more deeply into the intimate circle of knowing and loving set up by God between himself and us, that we are able to get a firm conceptual grasp of God, limited though it is, of such a kind that far from having God at the disposal of our formalised understanding of him we know him infinitely to transcend all our forms of thought and speech and worship. Thus the inadequacy of our theological concepts and statements belongs to the kind of precision they must have in truthful relation to him, and at the same time to the kind of spontaneous freedom with which God has endowed our personal and social being in created correspondence to his own. In this way theological science, working within the manifold complex of natural and human science, may carry its inquiry reverently to the point where the intrinsic intelligibility of the created universe is discerned to derive from and ultimately to repose in the uncreated Rationality and eternal Love of the Creator. It would not be very difficult to show through an analysis of the history of thought that the classical approach to the objective intelligibility of the universe which lies behind all our Western science and culture developed together with the Christian doctrine of the Trinity. It was the new and profound conception of the unity of God, as an internally constitutive unity, expressed in the doctrine of the Holy Trinity, which gave depth and solidity to the conception of the objective intelligibilities with which we operate in every area of human knowledge, art and religion.

It would not be very difficult to show either that whenever knowledge of God as a Trinity of Persons and a Communion of Love gives way to a conception of God in an undifferentiated oneness of Being, or where the doctrine of the one God is held apart from the doctrine of the triune God, there opens up a vast gulf between God and the created world in which God comes to be thought of as some sort of Unmoved Mover, and in which man for his part is unable to distinguish his own notions of deity from the Reality of God independent of his conceiving of him. In one case, the notion of God withers away as an otiose relic of an earlier habit of thought, and in the other case, room can be made for God only as a sort of hypostatised idea projected "out there", but to be reinterpreted in an oblique manner as merely a way of expressing our human attitude to existence. But further, by setting aside the doctrine of the Trinity on the ground that we cannot eradicate completely from our language expressing it certain anthropomorphic images which we do not wish to project into God, we deal a savage blow at the personal and intimate reciprocity that belongs to the very stuff of our personal relation to God, and we undermine the interpersonal structure of our understanding of him and communion with him. And so there becomes substituted for the concept of person, ontologically derived from the Holy Trinity, another concept of person, logically and impersonally derived, which is defined in terms of an individual's self-movement leading to a life of self-fulfilment and self-expression with all the ruinous individualism and selfishness bound up with it which are so painfully rampant in modern society.

I would venture to claim that it is to this introverted notion of the self that we owe so many of the psycho-pathological features of our social existence today. This replacement of the concept of the person arising out of the trinitarian structure of our relations with God by some sort of Boethian concept of the person, especially after its development through Lockean philosophy and romantic notions of self-consciousness, helps to account for the inability of countless people in the modern world to reach a convincing knowledge of the living God. Once we replace the trinitarian understanding of God by another with its simple mathematical unity, our minds are no longer able to get a proper grip upon his Reality, and we suffer from what Martin Buber called "a conceptual letting go of God".[50] Buber was not a trinitarian in his theology, but it would be difficult to challenge the contention that his basic concept of God owed not a little to Jesus whom he regarded as the greatest of the prophets, and to the Spinozan understanding of God as a movement of Love within himself which evidently reflects in an attenuated form the Christian doctrine of the Trinity as a Communion of Love in God.[51] It is in and through Jesus Christ alone that we have been able to reach an understanding of God in his inner divine relations which enables us to apprehend him in some measure as he is in himself and thus to be able to distinguish his transcendent Reality from the subjective states and conditions that arise in our knowing of him. This is the knowledge of God the Father through the incarnate Son and in the Holy Spirit which has the creative effect in those who live in communion with him of personalising and humanising their being.

NOTES

1. A. Einstein, see especially, *The World As I See It*, pp. 125f, 140, 153, 173f; *Out of My Later Years*, pp. 93f, 95ff; and the accounts of Einstein's conception of science by V. F. Lenzen and F. S. C. Northrop, in P. A. Schilpp, *Albert Einstein: Philosopher-Scientist*, pp. 357ff, and 387ff.
2. A. Einstein, *The World As I See It*, p. 174.
3. We cannot know God behind his back, as it were, by stealing knowledge of him, for we may know him only in accordance with the way he has actually taken in revealing himself to us. Hence, as Hilary has shown so well, we can only make use of the analogies and terms which God himself has provided and which he has linked to his self-revelation so that they point beyond themselves. It is in this way we interpret expressions like "father" and "son". *De Trinitate*, 1.19; 2.2–6; 4.14; 5.21; 6.13; 7.38; 12.9, etc.
4. The terms "father" and "son" can only be interpreted in a "sexist" sense if they are used as picturing models rather than as disclosure models. Used as disclosure models they function in divine revelation through "imageless" reference to God, in accordance with the second Commandment, Exodus 20.4.
5. St. Thomas Aquinas, *Commentary on the De Triniate of Boethius*, Q. 2, A 1 & 2. Cf. Karl Rahner on the problem of separate treatises, *De Deo Uno* and *De Deo Trino*, *The Trinity*, London, 1970, pp. 15ff.
6. A. Einstein, *Sidelights on Relativity*, "Geometry and Experience", London, 1922, pp. 27ff; and *Ideas and Opinions*, pp. 232ff.
7. L. Hodgson, *The Doctrine of the Trinity*, Edinburgh Croall Lectures, 1942–1943, London, 1943, pp. 89ff, 104f, 129, 175, 183ff.
8. Cf. H. Bouillard, *The Knowledge of God*, London, 1969, pp. 29ff.
9. Cf. again Bouillard, *op. cit.*, pp. 74f, 92f, in which he is deeply indebted to Barth and St. Anselm, although he wishes to reintroduce a notion of "precomprehension".
10. I have in mind St. Thomas' teaching both in *Summa Theologiae*, I a. 27–43, and *In Boethii de Trinitate*.
11. See especially Barth's sustained discussion in *Church Dogmatics*, 2.1, where he analyses the effects of a double approach to knowledge of God, of a detachment between the possibility and the actuality of knowledge of God, and the deep cleavage that arises in this way in an understanding of God's Being and his Act.
12. The principle with which St. Augustine operated here is that "whatever is said of God is said according to substance", for God *is* wholly in all his relations and attributes. His *esse* and his *existentia* coincide. *De Trinitate*, 7.5.10f.
13. Cf. Ephesians 3.14–15, where St. Paul makes the point that it is from the Fatherhood of God that all other fatherhood is named.

This is the epistemological principle of inversion that arises in knowledge of God, which Hilary expressed in asserting that all the analogies and expressions we have to use in theology are to be regarded as helpful to man rather than as descriptive of God, for they only suggest and do not exhaust what they indicate. *De Trinitate*, 1.19.
14. Cf. the account of St. Augustine's movement of thought given by J. N. D. Kelly, *Early Christian Doctrine*, London, 1958, pp. 276ff.
15. See at this point *De Trinitate* bk. 9 especially, where it is his thought of the self-giving of God in Word and Love which helps him to organise the trinitarian structures.
16. *De Trinitate*, 14.8.11.
17. *De Trinitate*, 14.6.4.
18. *De Trinitate*, 14.12.15.
19. Cf. *De Trinitate* 14.15.21: "But the mind is reminded that it may be turned to the Lord, as though to the very light by which it was in some way touched even when it was turned away from him"!
20. L. Hodgson, *op. cit.*, p. 153. See, for example, *De Trinitate*, 14.10.13, where St. Augustine seeks to show how a trinity is produced by the mind through remembering, understanding and loving itself. "And when the mind is thus turned to itself by thought there arises a trinity in which it is now possible to discern a word, conjoined both to the act of thought and its source through the will." Behind this lies the assumption that "knowledge is begotten of both, of the knower and the reality known" (*ab utroque enim notitia paritur, a cognoscente et a cognito*), *De Trinitate*, 9.12.18, which implies that "something of our own is subjoined to what we know", *De Trinitate*, 12.2.2.
21. This psychological slant of St. Augustine's thought, his stress upon self-consciousness or the in-turned habit of the soul, is most evident in *The Confessions*, but see also *De Trinitate*, 10.3.5, 5.7, 8.11, 9.12, etc. On the other hand, St. Augustine could insist that in self-contemplation the mind must not stay in itself but transcend itself in turning to God, *Sermons*, 143.7.9 and 330.3. Nevertheless St. Augustine's psychological slant had a damaging effect upon the history of western theology which laid it open to a constant interiorising of divine truth.
22. See the analysis by Viktor Warnach of St. Augustine's concept of the interior word, in "Erkennen und Sprechen bei Thomas von Aquin", in *Divus Thomas*, vol. 13, no. 3, 1937, pp. 272–289, in which he shows its influence upon the thought of St. Thomas.
23. *De Trinitate*, 15.23.43.
24. A. Seth Pringle-Pattison, *The Idea of God*, Oxford, 1917, p. 291. He refers in this connection also to A. Campbell Fraser, *Philosophy of Theism*, vol. 1, p. 77, and W. Windelband, *History of Philosophy*, Eng. tr., 1907, p. 238.

25. W. R. Sorley, *Moral Values and the Idea of God*, Cambridge, 1918, p. 299; cf. also p. 277.
26. W. R. Sorley, *ibid.*, pp. 123, 128ff, 498.
27. C. C. J. Webb, *God and Personality*, London, 1919, pp. 61ff.
28. C. C. J. Webb, *ibid.*, pp. 241ff; cf. pp. 128ff. Cf. in this respect St. Augustine's insistence that "for God it is not one thing to be, and another to be a person, but it is absolutely the same thing". *De Trinitate*, 7.6.11. Again: "We say three persons of the same essence, or three persons, one essence; but we do not speak of three persons out of the same essence, as though essence were one thing and person another." *De Trinitate*, 7.6.11. For St. Augustine, God is personal in that he is the Holy Trinity of Persons, but he warns against thinking of the Trinity *in* one God for he is himself God, *De Trinitate*, 15.23.43.
29. See R. J. Blaikie, *"Secular Christianity" and God Who Acts*, London, 1970.
30. Cf. the introverted cast of the *Soliloquia*, and not least the prayers, e.g., 2.1: "God always the same, let me know myself, let me know thee." (*Deus semper idem, noverim me, noverim te*); or 2.6: "Under thy guidance, let me return to myself and to thee." (*Te duce in me redeam et in te*).
31. See again Viktor Warnach, *op. cit.*, pp. 282ff. Cf. St. Augustine, *De Trinitate*, 12.15.24: "It ought rather to be believed that the intellectual mind is so constituted in its nature that by a kind of incoporeal light of a unique nature it may see those things which through the disposition of the Creator are subjoined by a natural order to intelligible realities, rather as the eye of the flesh sees things around it through a corporeal light for which it is made to have a natural capacity." St. Thomas, on the other hand, stressed the idea that our insight or judgment is due to a *created* participation in or reflection of the eternal Light of God, and thereby replaced the Augustinian idea of illumination by the light of the intellect. *In Boetii de Trinitate*, q.1, a 3 ad 1; *De veritate*, q.1, a 4 ad 5; q.10, a 6 c; q.84, a 5; q.88, a 3 ad 1, etc. For an excellent account of St. Thomas' position see Bernard Lonergan, *Verbum. Word and Idea in Aquinas*, London, 1968. For a defence of St. Augustine's view of enlightenment, see R. H. Lash, *The Light of the Mind*, Lexington, Kentucky, 1969, ch. 7, pp. 94ff.
32. Boethius, *De duabus naturis et una persona Christi, adversus Eutychen et Nestorium*, 2.1–5. "Wherefore if person is only in substances, and in rational substances at that, and every nature is a substance that exists not in universals but in individuals, the definition of person is found: an individual substance of rational nature." In the following chapter Boethius reduces the definition simply to "rational individual" (*rationabile individuum*). It is hardly surprising that Boethius could admit to a difficulty in thinking of

persons as other than one another in God, *De Trinitate*, 5. Cf. St. Thomas' handling of the Boethian definition, *Summa Theologiae*, I, q. 29, a 1–2.

33. See the analysis of Boethius' thought given by Heribert Mühlen, *Sein und Person nach Johannes Duns Scotus*, Werl/Westfahl, 1954, p. 2f. It was largely through Boethius' Commentary on the *Categories* of Aristotle, 2a–4b, that the distinction between primary and secondary substance came to occupy such a central place in the Latin and Mediaeval interpretation of Nicene and Chalcedonian theology, which has left a mark on Western Patristic scholarship still very evident today.

34. Contrast the balance of St. Augustine's words in *De Trinitate*, 7.4.7. "For the sake of speaking of ineffable things, that we may be able in some way to express what we can in no way utter fully, our Greek friends have spoken of one essence, three substances, but the Latins of one essence or substance and three persons ... Provided what is said is understood only in a mystery, such a way of speaking was acceptable, that there might be something to say in answer to the question as to what is meant by the three in the statements of faith ... The Sublimity of the Godhead surpasses the power of ordinary speech. For God is more truly thought than spoken, and exists more truly that he is thought." Cf. the reconciling effect of the *Quicunque vult*, or the so-called "Athanasian Creed", in the West with its emphasis upon the Unity in Trinity and the Trinity in Unity, or the similar effect of the *De sacrosancta Trinitate* of Pseudo-Cyril, incorporated in the *De fide orthodoxa* of John of Damascus, I.8, in the East.

35. Richard of St. Victor, *De Trinitate*, 4.22–24. See the text in Latin and French by Gaston Salet in the *Sources Chrétienne* series, no. 63, Richard de Saint-Victor. *La Trinité*, Paris, 1959, pp. 280–286. In application to God the definition becomes appropriately: "A divine Person is an incommunicable exsistence of divine Nature." While this concept of person was taken up by Alexander of Hales and developed by his pupil Bonaventura, and thus launched into the tradition of Franciscan theology, it was John Duns Scotus, also a Franciscan, who gave it the closest attention, *Opus Oxoniense*, 1.23.1. See also *Ordinatio*, 1 d.2, q. 1–4. Cf. again H. Mühlen, *op. cit.*, pp. 68ff, for an account of Duns Scotus' handling of the Ricardine concept of person.

36. See "Toward an Ecumenical Consensus on the Trinity", my account of a Colloquy held by the Académie Internationale des Sciences Religieuses, in March, 1975, based on Karl Rahner's essay, *The Trinity*, in which significant steps were taken in this direction, *Theologische Zeitschrift*, vol. 31, 1975, pp. 337–350.

37. Martin Buber, *Ich und Du*, Berlin, 1936, tr. by R. G. Smith, Edinburgh, 1937.

TRINITARIAN STRUCTURE OF THEOLOGY

38. I have in mind especially Martin Buber's American lectures of 1951, published as *The Eclipse of God*, New York, 1957.
39. M. Buber, *op. cit.*, p. 14f. Cf. D. Bonhoeffer, *Creation and Temptation*, London, 1966, p. 44: "The abstract concept of God is fundamentally much more anthropomorphic, just because it is not intended to be anthropomorphic, than childlike anthropomorphism."
40. M. Buber, *op. cit.*, pp. 14, 18.
41. Cf. Buber's critique of existentialist and psychological approaches toward healing people's schizoid conditions from a centre of unity within them instead of through a transcendent relation beyond them, *op. cit.*, ch. 5, especially, pp. 78–92.
42. The words of our Lord reported in St. Matthew, 11.25–27 and St. Luke, 10.21–22 about the mutual relation of knowing between the Son and the Father had a deep influence in the development of trinitarian theology in the Early Church. Cf. Athanasius' exposition in *In illud 'omnia'*.
43. See A. W. Wainwright, *The Trinity in the New Testament*, London, 1962.
44. The expression "mode of being" or "mode of existence" (*tropos hyparxeos*), which was widely used in Greek theology, derives from the Cappadocian theologians. See, for example, St. Basil of Caesarea, *De Spiritu Sancto*, 46, and Amphilochius of Iconium, *Fragment*, 15, Migne, *P.G.*, 39, p. 112. The term is to be understood in a strongly ontological or constitutive and *anti-modalist* sense. This is the way in which it has been used in modern times by Karl Barth, *Church Dogmatics*, I.1, Edinburgh, 1975 edition, p. viii.
45. This is already latent in St. Augustine's trinity of *amans, id quod amatur* and *amor* — cf. *De civitate Dei*, 11.26, and Karl Barth on this, *op. cit.*, p. 338.
46. *De Trinitate*, 5.11.2; 6.5.7; 15.27.50. A similar notion is found in St. Basil, *De Spiritu Sancto*, 13.30, 16.38, 17.43, 18.45f; Epiphanius, *Ancoratus*, 9–13; and John of Damascus, *De Fide Orthodoxa*, 1.13, which seems to refer back to Epiphanius, *Ancoratus*, 13.
47. This was the line taken by St. Athanasius in his epoch-making *Letters to Serapion on the Holy Spirit*. See the edition by C. R. B. Shapland, London, 1951. See my account of this in *Theology in Reconciliation*, London, 1975, pp. 231–239.
48. No one has taken this approach more effectively than Karl Barth, *Church Dogmatics*, 1.1, the section on "The Root of the Doctrine of the Trinity", pp. 304–333. Cf. the discussion of this by Claude Welch, *The Trinity in Contemporary Theology*, London, 1953, ch. vi, pp. 161ff.
49. D. Bonhoeffer, *op. cit.*, p. 39.
50. M. Buber, *op. cit.*, p. 125. What Buber steadily attacks is a merely abstract, conceptually empty notion of God, or "a purely negative

borderline concept" of God, p. 80.
51. See the propositions advanced by Spinoza about God's idea of himself and especially of his intellectual love of himself, *Ethics*, Part V, xxxii–xlii. This reads like an abstract version of aspects of St. Augustine's thought, but unlike his it is startlingly anthropomorphic in Bonhoeffer's sense. Cf. Buber's tantalisingly brief references to Spinoza's doctrine of *amor Dei, op. cit.*, pp. 14–20, and ch. 4 on "The Love of God and the Idea of Deity", pp. 47ff.